Fixed/Mobile Convergence and Beyond

Fixed/Mobile Convergence and Beyond

Unbounded Mobile Communications

Richard Watson

AMSTERDAM • BOSTON • HEIDELBERG • LONDON
NEW YORK • OXFORD • PARIS • SAN DIEGO
SAN FRANCISCO • SINGAPORE • SYDNEY • TOKYO

Newnes is an imprint of Elsevier

ELSEVIER

Newnes

Newnes is an imprint of Elsevier
30 Corporate Drive, Suite 400, Burlington, MA 01803, USA
Linacre House, Jordan Hill, Oxford OX2 8DP, UK

∞ Recognizing the importance of preserving what has been written, Elsevier prints its
books on acid-free paper whenever possible.

Library of Congress Cataloging-in-Publication Data
Application submitted

British Library Cataloguing-in-Publication Data
A catalogue record for this book is available from the British Library.

ISBN: 978-0-7506-8759-1

For information on all Newnes publications
visit our Web site at www.books.elsevier.com

Printed and bound by CPI Group (UK) Ltd, Croydon, CR0 4YY

Transferred to Digital Print 2011

Working together to grow
libraries in developing countries

www.elsevier.com | www.bookaid.org | www.sabre.org

ELSEVIER   BOOK AID International   Sabre Foundation

Contents

Foreword

Never before has the need to constantly stay connected been so great. It is for this reason in most industrialized countries, ownership of a personal cell phone is fast approaching market saturation. The compelling convenience of cellular phone use has progressed to the point where many – especially young adults, don't even have a fixed-line telephone but rely solely on a cellular phone for telephony services. Reliance on mobile devices is evidenced in business where analysts have noted that 40-50% of all business cell phone calls are made in sight of a desk phone. The convenience of mobility is too compelling to deny. However, cellular telephony alone does not meet all the mobile market requirements, due to limited in-building coverage and a lack of feature-rich businesses services.

There are benefits to using "fixed" telephony providers (traditional public switched telephone network, or PSTN, services) and "mobile" telephony providers (cellular services) but each also has its own set of drawbacks. Converging functionality provided by the traditional "fixed" networks with the mobility provided by the cellular networks is seen as the optimal solution. A widely-popularized solution to bridge this gap has been proposed by the telecommunications industry and termed *fixed/mobile convergence* or FMC. There are, however, many implementations of FMC coming to market that differ greatly in implementation and benefit realized while bearing the FMC label. These solutions range from PBX/IP-PBX add-ons to standalone solutions and service provider offerings. The vast array of disparate solutions has made it difficult to grasp the value of each and completely understand what FMC solution is best for each specific need.

At its core, FMC is a knitting together of multiple technologies (WiFi, VoIP, cellular, PBX, and Internet) and standards from multiple vendors, which further complicates understanding the scope and value of any one solution. The telecom professional has been faced with the challenge of learning about FMC solutions in a highly fragmented manner by reading publications, news websites, blogs, product datasheets, and white papers. There has been no single reference source available that describes how all these technologies are brought together, nor has there been any source that describes how these solutions are accessed through a sales channel. Filling this information gap was the motivation for writing this book and I know of few people as knowledgeable about this topic as the author, Rich Watson.

This book clarifies the morass of technical acronyms used to describe these emerging mobile communication product; it describes, in a straightforward manner, how each of the contributing technological elements adds to the total solution. This book is not intended to be a tutorial on each contributing technology, but rather has the goal of providing insight and understanding on how each element contributes to the overall FMC solution. Writing a book such as this is a challenge because of the rapid evolution of each the components but this rapid rate of change underscores the necessity of a single, unbiased resource for describing how FMC is implemented and how consumers and prosumers benefit.

The description of varied mobile communications solutions is found in this work along with an accurate annotation of the current state of products, key standards efforts and technology trends that will affect purchasing decisions for such products. Two basic FMC markets have evolved (consumer-centric and enterprise-centric), and solutions for these separate markets are addressed here.

This book fills a vacuum in the information space today regarding FMC, providing a full-spectrum description of the contributing technologies and the challenges and benefits of knitting each into an FMC solution that can be successful in the marketplace.

Rich Tehrani
President and Editor-in-Chief
Technology Marketing Corporation

Preface

What Is the Purpose of This Book?

The motivation for writing this book is to describe the emerging unbounded mobile communications (UMC) technology and market in a manner that is both tutorial and referential in nature, providing a knowledge base that couldn't exist until now, given the marked evolution that has taken place in the past five years. What UMC is and how it might be integrated into the consumer or enterprise ecosystem might be easily misunderstood or be confusing by simply reading the industry press. Simply stated, the purpose of this book is to provide a single source that will simultaneously educate both those responsible for mobile communication *buy* decisions and those charged with implementing mobile technologies with the knowledge to make sound decisions. Furthermore, I want to provide those responsible for making purchasing decisions with the sufficient market savvy to select the *best in class* or *best fit* for their business.

Why Is It Important?

Making the best decision regarding purchasing and deploying UMC solutions implies an assumption of knowledge about the functional benefits and corresponding costs of all the key solution components. Return-on-investment (ROI) assessments can be quite complex and somewhat subjective. A solid understanding of the technologies and market forces will aid in making the best decision aligned with customer needs.

In this book, a broad collection of alternative UMC solution approaches will be reviewed, along with the associated pros and cons. Each approach has specific value-added aspects that may be better suited for one particular market segment over another. This book will attempt to describe the details of the most mobile communication requirements for customer market segments as diverse as the consumer and enterprise markets. The stance taken in each case will be non-partisan, leaving the final assessment and purchase decision to the reader.

Who Is the Target Audience?

An underlying design of this book is to address two different classes of readers:

- *CFOs, CIOs, and IT managers.* Those who are responsible for making the final value-buy decisions and who do not need the details of the individual components and underlying technologies.

- *Network and telecom managers.* Those responsible for understanding the underlying technologies and how they might be implemented in addition to understanding the potential impact of certain configuration decisions.

Each chapter will be formatted to give a brief technology *tutorial* along with current market product trends and a statement about the status of the readiness and capability of that specific element technology to form solid UMC solutions. For example, it may be important to understand the state of any one UMC component's market *readiness* because it might affect the timing of a buy decision. Likewise, understanding some of the integration complexities involved in deploying a UMC system may assist in evaluating an SI or VAR proposal for such a solution.

How to Best Use the Information?

Each chapter covers a specific product or technology component of a total UMC solution. The beginning sections are directed to the buy decision makers. The balance of each chapter focuses on documenting the technical details sufficient to understand what is important to the success of a UMC deployment. These later sections are not intended to be comprehensive tutorials; rather, they are annotations of specific technology functional details describing how the technology impacts and contributes to UMC functionality. Full tutorials on WiFi, SIP, VoIP, telephony, or cellular networks may be found in other published works.

Attempting to write about a disruptive technology is problematic. *Change is constant.* During the writing of this book many new standards have been announced, new vendors have come into the market, new products have been introduced, and many company acquisitions have taken place. It is likely that some information in this book will be out of date at printing, despite all efforts to keep it current. To minimize any stale information, every effort has been made to ensure that all information is the most recent. The core technologies, however, are not anticipated to change significantly in the next 24–48 months, and the observations found in this book will be sound.

The hope is that with the knowledge derived from this work, UMC customers will be able to understand the market and the technology and make optimal decisions in purchasing and implementing unbounded mobile communication solutions.

Acknowledgments

Writing a book takes time. It is especially challenging when the subject you are writing about is in constant flux. Hours of thought and reading go into ensuring that what is articulated is best said to convey the exact ideas. The topic of UMC is particularly challenging because it requires integration of so many diverse technologies to bring a unified solution to the market. The evolution of our social structures demands greater freedom in communication options. Proliferation of wireless technologies becomes the basis for that freedom—a freedom without geographic bounds.

Because of the extensive span of different technologies of UMC solutions, it is difficult for one person to fully grasp all the details of each contributing element. It takes input and critique from specialists in the individual areas to ensure that the message is on target. I am indebted to the following friends and professional comrades for their time and valuable input to ensure that the content of this work is accurate:

- Clint Chaplin, chairman of the IEEE 802.11r Task Work Group and past chairman of the WiFi Alliance, Mountain View, CA

- Steve Shaw, VP of Marketing for Kineto Systems, Milpitas, CA

- Jenni Adair, Director of PR for DiVitas Networks, past Director of PR for Trapeze Networks, Mountain View, CA

- Mark Ferrone, PR Manager, Customer Programs, Corporate Communications for Cisco Systems, Santa Clara, CA

- Jeff Watson, VP of New Media, Warner Bros Records, Burbank, CA

- Amanda Mitchell Henry, Former editor of *InfoWorld* (San Francisco), *LAN Times*, and *Computer Reseller News*, now a technology industry freelance writer

- Bob Beach, Senior Director of Engineering, Motorola Enterprise Division, San Jose, CA

- Bob O'Hara, Co-founder of AireSpace, Inc., and Director of Systems Engineering – retired, San Jose, CA

- Dave Hockenberry, Senior Technologist for Verizon, Mountain View, CA

- Barbara Nelson, CTO of iPASS, Inc., Redwood Estates, CA

- TJ Noto, Director of Business Development, Boingo, Inc., Los Angeles, CA

- Marc Solsona, Director of FMC handset development for DiVitas Networks

- Nora Freeman, Senior Research Analyst, Enterprise Networking for IDC

In today's ultra-high-tech world, it takes multiple perspectives to grasp the full scope of the UMC solution's complexity. To reach the goal set for this work takes the collaboration of a unique team of individuals contributing their learning and insight. As the late tennis pro Althea Gibson observed, "No matter what accomplishments you make, somebody helped you." Thank you all!

I will always be grateful to my wife, Geri, for her patience and editing help in the process of writing this book.

I believe a book on this topic, with its overview perspective and its target of assisting the mobile market decision makers in understanding UMC solutions and making the best product selection, is timely. I trust the book meets those goals.

CHAPTER 1

Unbounded Mobile Communications

1.1   Communication Knits Societies Together

When the Minneapolis I-35W Bridge collapsed on August 1, 2007, it couldn't have happened at a worse time. It was the middle of the evening commute and untold numbers of cars and trucks were on the bridge when it went down in those fateful few seconds. Not only were the massive bridge's roadway parts in the Mississippi River, but hundreds of people were struggling for survival in the chaos. With the bridge collapse, most of the communication links were also severed, hampering rescue efforts; wireless services were the only remaining communication links still operative. As the rescue teams launched their efforts, their communications relied solely on the wireless services from cellular and WiFi networks that covered the bridge area. Quickly, voice links were established over the cellular network, and because of the proximity of the municipal WiFi service, Web cameras were set up to continually monitor the site and to aid rescuers in focusing their efforts. The wireless communication services in place helped save lives and minimize the trauma of this disaster.

Communication among people has always been at the cornerstone of success for all civilizations; whether spoken, written, read, viewed or heard, it is how we progress, learn, develop, adapt, express, and pass on knowledge, faith, wisdom, and history. Whether by the cave drawings of early humans, Native American smoke signals, the Gutenberg press, or the intergalactic radio probes of the 21st century, these different forms of communicating ideas, concepts, and information have been the basis of how we have progressed. However, it is not only *what* we communicate, but *how* what we communicate impacts each successive generation and the means by which we do it.

In the 21st century we take for granted the presence of communication services, whether television, home/office phone, pager, or cellular phone. Each successive generation has adopted the latest communication technologies and abandoned the technologies of the past (remember teletype or telegraph or pagers?). The ability to reach out and "touch" someone is a cultural assumption, and industrialized nations feed on a constant stream of information. The major trend sweeping our cultures in the past 30 years has been wireless communications. As individuals, we have become more mobile throughout our daily happenings, and communication between any two people has to accommodate this mobility.

There are roughly 291 million wireless cell phone subscribers in the United States, which now has an estimated population of about 301 million.[1] Worldwide, the adoption of cellular phone subscribers is over 80% in developed nations and approaching 50% for all countries, meaning that some 3 billion cell phones are in daily use on the planet. This is a clear indication that today's communication method of choice is *wireless*. Other statistics indicate that upward of 8% of North American households[2] no longer have a landline phone and use only a wireless phone as their primary method of communication. Adding to these data is information that the average individual in the business sector carries more than two mobile devices (cellular phone, personal digital assistant [PDA], iPod, iPhone, laptop computer, or the like) as a matter of daily work. The trend is clear: *Wireless communication is important to all urban societies around the world.*

One fact dominates the modern world: We are a mobile society, rarely stationary long enough to communicate from a static phone connected to a wall or on a desk. In earlier times, people often accepted missed calls as a fact of life. Today voicemail is no longer a nice-to-have option but an assumed function. Missing a call to someone, we usually expect to be routed to their voicemail to leave a message with the hope that at some later time they will return the call.

"Back in the old days," if both parties were away from their desk phones, the proverbial telephone-tag ensued. Communicating via cellular phone minimizes this problem but has problems of its own: inadequate coverage. Early in the history of wireless phone service, relatively small geographic zones of large metropolitan cities had cellular phone service. You could make a call while downtown, but if you

---

[1] Clearly, the ratio of these numbers indicates that some individuals have multiple wireless devices. Either that or there are a number of elementary and preschool children who also have their own cellular phones.
[2] CTIA, 2006.

traveled outside of town, the call would most likely drop due to a lack of cellular coverage.[3] As the popularity of cellular phones grew, the business justification for expanding wireless network coverage was clear. Most of us now naturally assume cell coverage in most populated areas. Yet calls are still missed, calls are dropped, and telephone-tag and voicemail are still with us.

1.2 The Business Value of Mobility

Even in the 21st century, with third-generation (3G) cellular technology being deployed, cellular communication does not meet all the desired requirements. Coverage is still a problem. In rural areas, cellular coverage might not exist. The major urban problem with cellular coverage is the fact that it might not penetrate into buildings. Brick, stone, plywood, and steel are opaque barriers to the cellular RF signals. In many offices today, spotty cellular coverage may be available, but many times the astute caller may need to move near an outside window to obtain a good cellular signal to make a call. Though this will accommodate outbound calls, it does little to enable a cell phone user to receive inbound calls that were missed while inside the building. For such calls, voicemail is the last resort and must be checked periodically. The desire to use the wireless service is so prevalent that some studies have shown that up to 40% of cellular phone calls (mobility is driving this issue) are made inside the office within sight of a desktop phone.[4] However, it is a simple fact of life that many homes, offices, and public buildings can be virtual cellular dead zones. And in these dead zones, the mobile worker is frustrated because of the broken communication link. Coverage is the number-one requirement for the mobile user.

The desire of the savvy mobile public is to have communications available *everywhere*. The goal of the telephony vendors is to meet this demand because it not only stabilizes but expands their subscriber base, creating new value-add business opportunities. The business forces that come into play in meeting these mobility requirements are varied. Traditional landline vendors (providers of the traditional wired desk and home phones) see their business base dwindling as more customers move to a pure wireless solution. Cellular carriers see an opportunity to expand their customer base and build loyalty with their subscribers, slowing or preventing subscriber *churn* (customers switching to a

---

[3] Even in urban areas, forecasting capacity requirements is a challenge that calls for spending funds to expand coverage before the customer demand is realized.

[4] On average, mobile calls make up more than 40% of all calls made or received on the job, according to IDC.

competing service). New, disruptive technologies like Voice-over-IP (VoIP) and Wireless LAN (IEEE 802.11, known as WiFi) have created opportunities for new competitors to come into the telephony market and vie for these same mobile customers. The way each of these vendors chooses to meet this market need is centered on the concept of how to bridge calls that were traditionally directed to a desk or home phone (a fixed line) with cellular (wireless) services and how to make mobile access more pervasive. Solutions targeted to meet these evolving communication requirements are often labeled with the term *fixed/mobile convergence* (FMC). Delivery of communications solutions with such sophistication of technology will result in *unbounded mobile communications* (UMC) by virtually eliminating the communication dependencies of geography or wireless access types. Throughout the remainder of this book, references to *unbounded* communications capabilities will use the *UMC* acronym. Though there are branded mobile solutions that use other acronyms and popular terms that are defined in subsequent chapters, UMC encompasses the amalgamation of multiple technologies that work together, resulting in a seamless communications solution.

1.3   Unbounded Mobile Communication Concepts

How does UMC solve mobility problems? The concept is that new wireless technologies will be "married" and result in solutions that provide virtual ubiquitous wireless access for communications, regardless of the user's specific geographic location or proximity to cellular coverage. A number of emerging wireless technologies can "fill" the spaces not covered by the wide area cellular networks, and wireless LAN WiFi is becoming pervasive in home and office, a natural candidate for the marriage. Such possible solutions would combine the utilization of WiFi for in-building (or on-campus) wireless coverage with access to existing wide area cellular coverage (see Figure 1.1). In this manner, whether a person is in a building or outdoors, he or she can remain connected to a virtual network and make and receive calls as well as run applications over these networks transparently.

Additionally, a major feature of a UMC solution is the ability to seamlessly transition between disparate wireless networks (WiFi & cellular) without dropping a call. To construct a UMC solution, multiple technologies must be knitted together and presented as a *solution* to the consuming public. How are these technologies linked together to form viable solutions? The following chapters detail the various UMC solutions with a brief tutorial and history on each individual technology along with details on the

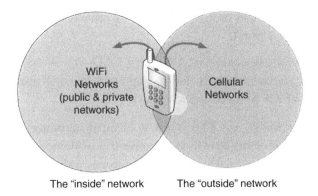

The "inside" network   The "outside" network

Figure 1.1: Seamless roaming across wireless networks.

implementation challenges that are unique to each technology element in addressing the overall UMC opportunity.

The term *FMC* is found everywhere in popular technical and industry journals and papers. This label has been somewhat overused and even misused to the point that many are confused as to exactly what is being described. There are also additional terms emerging to define similar products that have slightly different architectures and that also add to the complexity of understanding the entire solution set. The UMC term is used in this book to frame a class of products that is inclusive of those with an *FMC* marketing label but also describing other mobile communication approaches. A clear definition of the different usages of FMC and other terms is given in a subsequent chapter.

1.4   UMC: What Is Needed?

Any UMC implementation will depend on the availability of all or part of the following technologies, products, or services (see Figure 1.2):

- *Dual-mode handset.* A single mobile device that has been designed to operate over both with 802.11/WiFi and cellular (GSM or CDMA) wireless networks.

- *Wireless LAN (WLAN) infrastructure.* An IEEE 802.11 wireless service that is configured and available to support IP-level connectivity between key UMC elements within the network. An example would be a WLAN located at corporate headquarters or a WiFi network available from a mobile worker's home or public hotspot.

- *Wireless carrier infrastructure.* GSM or CDMA (or other) coverage. Use of a cellular data packet service is also an important carrier feature that opens access to the Intranet for IP-level application support.

- *A mobility service/server.* This is a new element introduced with the support of UMC that manages the transition ("hand-off") of devices as they switch between WiFi and cellular networks and ensures call continuity. This can take the form of one or more server components installed in a business or within the carrier "cloud."

- *Application services/server.* For businesses, one UMC benefit is being able to connect with key applications across multiple wireless domains. The premier UMC service is a connection to a Private Branch Exchange (PBX). This implies interface with PBX systems and support for popular telephony application features such as call transfer, call hold, call conference, and message-waiting indication.

- *Voice over IP (VoIP) services.* Access and deployment of VoIP is not crucial but is often coupled with UMC implementations. Specifically, the Internet Engineering Task Force (IETF) Session Initiation Protocol (SIP) is the international standard that seems to be the predominant standard for most VoIP providers.

- *IP Multimedia Services (IMS).* This proposed network service class (based on IP and SIP) will be a major force in accelerating deployment of not only wireless mobile communications but also worldwide distributed networking for virtually all nations, individuals, and commercial businesses alike. IMS provides a network platform for supporting a virtually unlimited set of applications, including UMC. IMS-supporting products, promoted by cellular service providers, will be on the market by late 2008 and are expected to mature over the next three to five years.

Each of these UMC solution components may still be in its own respective state of evolution toward maturity. Additionally, no single vendor has announced a complete end-to-end UMC solution, which makes delivery of a full solution a true channel marketing challenge. Some components are also optional for enterprise UMC solutions (dotted line in Figure 1.2).

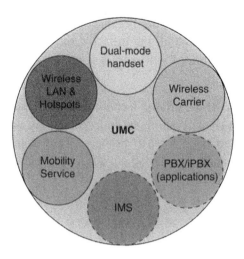

Figure 1.2: UMC solution components.

1.5 Are the Technologies Ready?

Are UMC product development and market channel delivery problems insurmountable? The simple answer is no. The challenges facing the UMC market are very much like those with consumer products (e.g., the automobile, television, or the cell phone). Each of these technologies was "disruptive" and radically changed the way things were *done*. In the relatively recent past, people began to ride cars instead of horses. Later they watched live images on a screen instead of listening to radio. And then people began to talk on mobile phones, untethered by copper wire in the kitchen or living room. Each of these technologies was eventually embraced by the public after they became available. However, this did not occur overnight, because all these technologies needed a requisite support infrastructure to reach market maturity.

For the automobile, it took more than 150 years of development to go from concept to mass production at a price point for the masses. Even then, use of the car was limited because there was no nationwide gasoline production, no gas stations, no trained service people, or most of all, no highway system. In fact, in the United States, it took another 50 years to have the highway systems in place before the average car owner had the freedom to drive across the country without the fear of being stranded without gas or blocked from proceeding because no roads were leading to their destination. Full adoption required all the infrastructure elements to be in place. The deployment of UMC solutions will require a similar level of mature infrastructure development and sophistication.

With specific reference to the underlying technologies required for UMC, each has been on an evolutionary path that will culminate in a technology intersection that will provide an infrastructure to support UMC products. Each solution element needs to be functionally complete and customer adopted before UMC products can be successfully marketed.

1.5.1   Cellular Phone History

In many countries, the per-capita population of cell phones is approaching saturation (i.e., every potential customer has purchased a cell phone). Like all disruptive technologies, general adoption of the cellular phone did not happen overnight. In fact, it took over 50 years of technological development, infrastructure deployment and social acceptance for this to be realized.

As indicated in Figure 1.3, the history of the cellular phone business is annotated by a "generation" terminology. It has taken more than 30 years to go from the "zero" generation (0G) to the third generation (3G). Even with this lineage, the cellular technology is still being enhanced with a 4G specification being defined. With respect to UMC, the features supported by the 3G networks are a perfect match and are not a hurdle for UMC deployment.

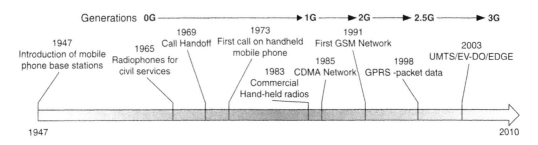

Figure 1.3: Cell phone product milestones.

1.5.2   Wireless LAN History

Whereas the wireless local area network (WLAN) business had its genesis much later in the 20th century, the product was not "birthed" fully matured. Figure 1.4 presents a timeline of the evolution of WLANs. Beginning in the 1970s, a core wireless technology was developed to meet a need for communicating between the Hawaiian Islands, thus the name *AlohaNet*. As LANs became more popular, it soon became apparent that there was a need to be connected with the LAN without being tethered by the Ethernet cable. To meet this need, proprietary WLAN products began to appear on the market in the late 1980s (Proxim and Symbol Technologies). Adoption of this new technology was restricted to some vertical markets (e.g., retail and healthcare) where the "pain point" caused by not being mobile was the greatest. However, the market began to take notice of the benefits a WLAN provides. In the early 1990s, the IEEE formed the 802.11 working group to define an international standard for a wireless LAN. A significant milestone was reached in 1997 with the ratification of the 802.11 WLAN standards, which has resulted in the creation of a multibillion-dollar worldwide business. Today, as you might know, the WLAN market is dominated by several main players, including Aruba, Cisco (with Aironet/Airespace), Trapeze, Meru Networks, and Motorola (Symbol Technologies).

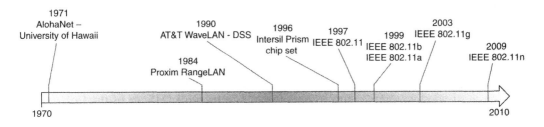

Figure 1.4: WLAN milestones.

As monumental as it was, the first 802.11 standard was weak in several key areas, especially security and quality of service (QoS), which slowed its adoption in the enterprise. Since 1997, the IEEE committees continued to extend and enhance the base set of standards to allow for development of more secure, more reliable, and faster WLAN products. There are still works in progress with regard to support of optimized wireless VoIP and QoS.

1.5.3   VoIP and iPBX History

The next two UMC functional requirements are linked tightly together. VoIP is the method by which voice is transmitted over a packet-switched network. This is radically different from the traditional circuit-switched networks of the legacy telephony public switched telephone network (PSTN). Like most emerging technologies, many VoIP products came to the market as proprietary solutions with no intervendor interoperability. In the mid-1990s, a number of international standards bodies launched study groups to define protocols for supporting telephony functions over the intranet. Most significant were the ITU H.323 and IETF RFC 3261/SIP efforts. Out of this work came competing VoIP standards that were quickly implemented as commercial products. Similarly to a market share battle between Beta-Max and emerging VHS videotape technologies in the consumer space, these two VoIP technologies were vying for market dominance. It appears from the sheer magnitude of the market adoption that SIP has become the VoIP standard of choice for the world.

By the turn of the 21st century, all major PBX manufacturers had acknowledged that VoIP was the telephony technology of the future, and each had announced development of VoIP-based PBX systems known as IP-PBX (iPBX), as shown in Figure 1.5. Whether offered as a standalone IP-only product or converged as a TDM-VoIP PBX hybrid, all work on the old analog or digital circuit switched products virtually came to a halt. Eventually, only iPBX systems will be offered commercially and will be deployed in a majority of enterprises in the next three years (see Figure 1.6).

There have been a large number of IMS-based product announcements. However, though most vendors today claim some interoperability with IMS, as of the first quarter

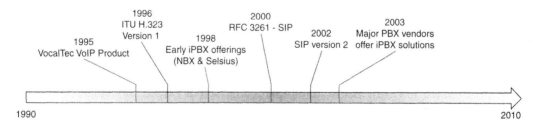

Figure 1.5: VoIP & iPBX market milestones.

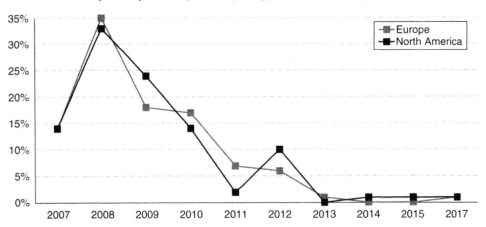

Figure 1.6: Enterprise adoption of iPBX products.

of 2008 no IMS products were broadly deployed. It is important to note that today, IMS is expressed as an *architecture* and *not* a specification. Therefore, there may be many incompatibilities in attempting to interconnect disparate vendors' IMS-*compatible* products. The IMS Forum and Global MSF, similar to the WiFi Alliance, have been formed for the purpose of education and interoperability validation between IMS-compatible products.

1.5.4　The Intersection of the Right Technologies

By coincidence or design, it seems that all the major technology components necessary to deploy a UMC solution are commercially available. The wireless components (WWAN and WLAN) are now commercially available with the right functional mix. VoIP has been adopted by the major players in the telephony market. The following chapters will discuss the state of the various contributing technologies and where their value-add is applied to a UMC solution. Are the other barriers in delivering UMC solutions? Yes, one of which is finding effective market delivery systems; more about this in later chapters.

1.6   What UMC Market Forces Are at Work?

As with any disruptive technology, there will be competitive forces in the marketplace to hinder or block market progress. This is true with any UMC solution.

Major market forces that affect how UMC products will be offered will impact vendors that fall in two general classes:

- Those who stand to lose business if UMC becomes popular

- Those who want to dominate the UMC market as the leading vendor

With UMC, the vendors who stand to lose the most market share are the traditional wireline carriers (the PSTN), which are those vendors who have brought you the dependable desk phone. They have already lost much of their traditional customer base to the cellular services. Many younger people no longer have a home landline but rely on telephone access via a cellular phone. This is also true of companies like Ford Motor Corporation and Nokia, where thousands of their own associates have no desk phone at all and depend solely on cellular voice services. To compete, fixed wireline service providers have had to change their business tactics and (1) offer more features and services as part of their product (i.e., voice and data over the same line), and (2) have begun to offer some wireless service complementary with their traditional service offerings. Technologies such as VoIP/UMC threaten to replace the aging wireline technology with faster, cheaper, and more mobile products.

Wireless (cellular) carriers also are similarly threatened because they will lose subscribers to traffic diverted over the Internet. Wireless carriers already combat increasing margin pressure from peer competition. This occurs, for example, with *churn*, which happens when customers continually switch from carrier to carrier to take advantage of some new feature or pricing program. The average revenue per user (ARPU) has declined and carriers are concerned that UMC type products will further erode their margin base.

Like the California gold rush of 1849, new companies have spawned and rushed in to claim part of the exploding UMC markets. Coincidentally, large established companies have also evaluated the business opportunity posed by this UMC concept, and they have added themselves to the number of vendors vying for a place in the market. Today, more than 90 vendors have announced their intent to offer some or all the solution components required by an UMC solution. Cisco has expanded its traditional network infrastructure offerings to now include VoIP-based iPBX products (Cisco Unified Communications Manager[5]), and the networking giant works with Nokia and other handset manufacturers to offer mobile phone solutions.

Each of these players has an agenda with regard to support of UMC. Some will implement a delay tactic because of potential loss of their current business base. In doing so, they will make it difficult to implement and deploy such solutions. Others will be aggressive in embracing the opportunity, even changing the basic technology of their companies to follow this new trend. As each vendor's strategy adds to the momentum of the market, it will become a river that may become treacherous. Not all will survive. Such forces will impact how fast and how rich a product set will be delivered to the consumer over the next five to 10 years.

Because a UMC solution requires functional integration of diverse technology components from multiple vendors, this implies that any successful UMC solution will be the result of a collaboration or convergence of multiple channel and technology partners. Strong product integration and marketing partnerships will be the hallmark of UMC successes, and both consumer and prosumer will be the ultimate winners.

1.7   Convergence in the Market

The word used to characterize the UMC functional intersection between wireless services is *convergence*. At the product level, multiple radio technologies are *converged* onto multifunctioned devices that support seamless roaming across divergent wireless topologies. This seamless access is also a functional convergence of the two wireless network services, which now appear to be a single service without boundaries.

The state of flux in the today's market has demonstrated that a *convergence* is happening, even at the UMC-component vendor level. No one vendor has had the technology breadth to supply all components of a UMC solution. Therefore, certain mergers and business collaborative efforts have occurred over the past 18 months that clearly indicate that many major market players want to extend their technology breadth to encompass more of the whole UMC solution (see Table 1.1).

---

[5] Cisco Unified Communications Manager (formerly Cisco Unified Call Manager) is the heir to the Selsius Systems legacy. Cisco bought Selsius in the late 1990s as its entrée to the VoIP market.

Table 1.1: Corporate convergence of businesses

Mobile market vendor	Business event	Assessment
Cisco	Purchase of Airespace	Extends its WiFi product offering
	Purchase of Orative	Adds "presence" service capabilities
	Purchase of WebEx	Extends online collaborative capabilities
	Unified Communications Manager (v6.0) extends desktop to smartphone	Extends VoIP solutions to mobile cellular phones
Avaya[6]	Purchase of Traverse Networks	Extends foreign voicemail services to mobile devices
Siemens/Nokia	Nokia-Siemens-Networks (NSN) joint venture	Partnered to offer a "solution" to the market
Polycom	Purchase of SpectraLink	Premier WiFi-only business for enterprises
CounterPath	Acquisition of FirstHand and Bridgeport Networks	Adds FMC capabilities (FirstHand & Bridgeport) to its strong SIP-based workstation softphone product line; Bridgeport initiated the VCC architectural design for 3G and is focused on IMS centric deployments
Research in Motion (RIM)	Purchase of Ascendent	Adds PDA and smartphone PBX mobility to Blackberry offerings
AT&T	AT&T Unity (Purchase of Cingular)	Economically converges the companies' fixed and wireless networks and allows unlimited calling between the two networks for common subscribers
Motorola	Purchase of Symbol Technologies	Enter the mobile terminal market with rugged dual-mode and WiFi WLAN product family
Comcast/Sprint	Collaboration	Pivot: Home-to-mobile service agreement links fixed home phone with mobile phone

---

[6] Avaya was taken private through a buy out in June 2007 by TPG Capital LLP and Silver Lake Partners.

Beyond the union of products through company mergers, a number of strategic alliances have been announced in which companies plan to work together and deliver UMC-like products. For example, in late 2007, Alcatel-Lucent (*AlcaLu*) announced its FMC strategy in collaborating with Aruba Networks, which it said would result in an integrated WiFi and iPBX product sold through one channel. Similarly, Nortel, 3Com, and NEC have announced collaboration with FirstHand Technologies to create an FMC product offering. Other such alliances will be made public to bring the proper product and support elements together in offering a UMC solution.

CHAPTER 2

Mobile Communications: State of the Technology

It was always amazing to watch the adventures of the *Star Trek* team and see Captain Kirk stranded on a planet or on another spaceship and yet able to speak directly with his compatriots via his "communicator." Not once did he use a keypad to *dial* Scotty, Spock, or Bone's phone number, but he rather simply spoke directly into the small device and was instantly connected to them. How did the device know to whom to connect—digital ESP? What was the range of this wireless wonder? Whatever it was, it represents a high benchmark for the ultimate in mobile communication: wireless interplanetary communications—true unbounded mobile communications!

We all acknowledge this as pure fiction, but the dream of UMC has grown from a *nice-to-have* to a *must-have* requirement for many businesses and nomadic individuals. Being mobile (away from the desk or home) is the norm today. The chance of catching someone at his or her desk is becoming more problematic, and telephone-tag inhibits fluid interpersonal communication. The use of cellular phones to meet general mobility requirements seems to be a partial answer, but it only addresses one part of the overall requirements: *off-campus or outdoor connectivity*. Often, such mobile solutions only serve to compound communications challenges by creating the need to continually check both cell-phone-based and corporate-based voicemail while out of the office. Also, cellular coverage is often too spotty to be considered reliable from a business perspective. It is not uncommon for a business user to step inside a building and instantly lose connectivity. Additionally, for business uses, standard cell phones provide no coupling with corporate information systems such as private branch exchange (PBX), instant messaging, vertical market voice applications, and email.

UMC is only one way to express the mobile connectivity nirvana. In this book, UMC is used as an umbrella term for a number of offerings in this market. A number of industry terms and acronyms have been created to describe different mobile communication solutions: fixed/mobile convergence (FMC), enterprise FMC (eFMC), Universal Mobile Access (UMA), Voice Call Continuity (VCC),[1] Generic Access Network (GAN), cellular-broadband convergence (CBC), Mobile Unified Communications, seamless mobile collaboration (SMC), and mobile-to-mobile convergence (MMC)—a confusing array of mobile solution labels and acronyms!

What are these technologies? How are they different? Do they mean the same things? And what do they mean to me? These questions are being raised more frequently in business periodicals, technology journals, and even articles in the consumer press. The subject of wireless mobility and convergence is becoming popular in a broad spectrum of publications. For those who live in the technology journalistic storm, the frequency of articles written on this topic makes it increasingly difficult to keep up with the latest industry news. These acronyms are often bound tightly to other topics such as Voice-over-IP (VoIP), 802.11 Wireless LAN (WiFi), 802.16 (WiMAX), and advanced cellular technologies such as Evolution-Data Optimized (EV-DO), General Packet Radio Service (GPRS), and Universal Mobile Telecommunication System (UMTS). The core attraction of the UMC concept remains the ability to roam, unrestricted, back and forth between public and private wireless networks, without user regard for connectivity requirements.

To understand how UMC services and product may be developed and deployed, it is important to understand where we are today with respect to the available mobile communication options. Principally, there are two major classes of wireless communications options:

- *Outdoor wireless.* The wide area wireless networks or cellular wireless.

- *Indoor wireless.* WiFi (via WLAN).

The former has had the greatest impact on mobile communications for consumers; the latter is becoming more important to businesses, thanks to quickly emerging UMC solutions. This chapter reviews critical important mobile communication features, status of commercial availability, and general challenges to UMC.

---

[1] From the 3GPP v6 standard.

2.1   Wide Area Wireless

Almost everyone on the planet owns a cellular phone, or that's how it seems.[2] Cellular wireless has evolved over the past 20 years and is so prevalent that most people—both consumers and business users—get frustrated when service is not available or if it drops their calls. All cellular users have experienced poor voice quality and a dropped call at one time or another. Worst yet is the frustration due to lack of coverage inside buildings! It seems that many business phone calls are attempted while inside a building with limited coverage, and thus mobile workers must be near an outside window or they must exit the building to find sufficient cellular coverage to make the call. As frustrating as it sounds, this is a common scenario in the workplace. Contrast this fact with the Pyramid Research (2005) observation that found "that up to 50 percent of wireless minutes are conducted from inside your business." This means that inside a building, even with potential marginal coverage, subscribers prefer accessibility provided by their mobile cell phones over the reliability of stationary desk phones. This trend has gained such momentum that some pundits have predicted the future demise of the office desk phone.[3]

The cellular carriers have spent billions of dollars worldwide to roll out their networks to provide the broadest cell coverage possible. Initially working from the most densely populated urban areas, they have progressively extended the coverage to less populated areas. The dominant base technologies of the worldwide wireless cellular networks fall into two categories:

- *Code Division Multiple Access (CDMA).* An early wireless technology, still dominant in North America and Asia/Pacific countries.[4]

- *Global System for Mobile Communication (GSM).* The most popular wireless standard in the world.

Both of these technologies have evolved extensively over the past 20 years through the addition of new features (voicemail, three-way calling, and the like), extended cell coverage, text messaging, and IP-packet services. The following sections provide a status report on these cellular technologies.

---

[2] For example, the U.K. has approximately 71 million cellular subscribers, with an estimated population of only 60 million (ABI Research, August 2007).

[3] "Will enterprises hang up on desk phones?" *InfoWorld*, May 25, 2007, Stephen Lawson, IDG News Service; www.infoworld.com/article/07/05/25/Will-enterprises-hang-up-on-desk-phones_1.html.

[4] CMDA worldwide subscriber distribution: North America, 33.6%; Asia/Pacific, 46%; Latin America, 17.9%; Europe, 2.5% (The CDMA Development Group, 1Q-07 report).

2.1.1 GSM Overview

By far GSM is the most popular cellular service worldwide, with more than 2.5 billion subscribers in 2007 and estimates of over 3 billion by the end of 2008.[5] Over 80% of the global wireless subscribers are connected over GSM networks. The basic network architecture of a GSM network is fairly straightforward and consists of (see Figure 2.1):

- *GSM handset.* A phone or mobile device designed with a GSM radio for cellular connectivity. The identity of the user is written in the Subscriber Identification Module (SIM) to which the service agreement is bound. Phones may be "locked" to a particular cellular provider through agreements with the manufacturer. This practice was widespread in the early years of GSM. However, many sources of "unlocked" phones are becoming available, most notably in Europe vs. the United States, which allows for a European GSM mobile phone to be used on any GSM carrier service once there is a valid service-level agreement (SLA) in place.

- *Cellular base station.* This is the static transmit/receive portion of the network, and it is strategically located throughout an area to provide the radio frequency (RF) coverage necessary to support cellular traffic. In some urban areas, base stations may be camouflaged as "trees" or advertising sign frames to hide their antennas. The handset communicates with the base station when mobile and will roam between geographically placed base stations as they move through an area.

- *Mobile switching center (MSC).* This key element in the network acts as a consolidator and command point to direct traffic (both voice and data) in and out of the network. In a metropolitan area, a highly mobile user might not only roam from base station to base station, he or she could also be associated with multiple MSCs during one connection session. An MSC "talks" to another MSC to hand off connections between them as users traverse a wireless coverage area. All this is transparent to the mobile phone end user.

- *Mobile Switching Center Gateway.* To make a phone call to a land line (non-IP) phone, the call may be directed to a MCS gateway that manages access to the PSTN.

---

[5] 3G Americas (www.3gamericas.org), "More than a million new users daily," www.3gamericas.org/English/News_Room/DisplayPressRelease.cfm?id=2982&s=ENG.

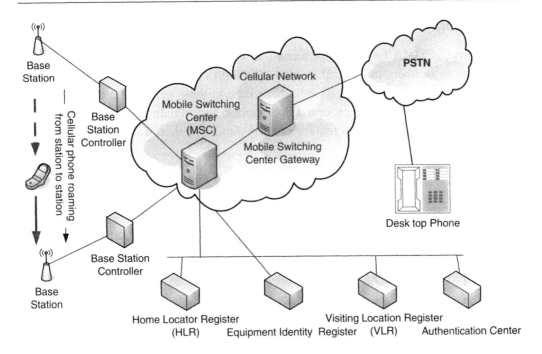

Figure 2.1: GSM cellular network components.

There are a number of service support elements within the GSM network, including:

- *Home Locator Register (HLR).* This is a very important component in the network because it is the repository of all the authenticated user information for that service provider's customer base. You cannot make a call on that network without being authenticated by a service on the HLR.

- *Visiting Location Register (VLR).* This element provides temporal access information for subscribers who might not be in the HLR but who are roaming away from their "home" network and the two networks have a roaming agreement.

- *Equipment Identity Register (IER).* This element tracks the authenticated devices that have contracts to access this network.

- *Authentication Center.* Performs Authenticate, Authorize, and Account (AAA) for all users accessing the system.

2.1.2 CDMA Overview

By popularity, CDMA networks are second in worldwide deployment with almost 400 million subscribers in 2007. Principally in North America and Asia/Pacific, this cellular class continues to grow its subscriber base in these geographic areas. Though using different radio frequencies, encoding schemes, and signaling protocols than GSM, CDMA's network component structure is fairly similar to that of a GSM network (see Figure 2.2).

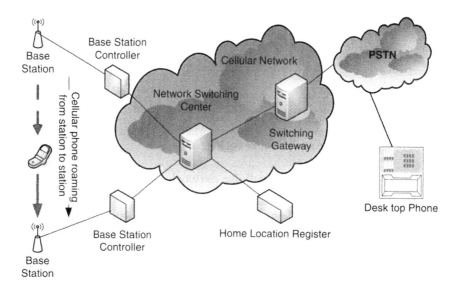

Figure 2.2: CDMA network components.

- *CDMA handset.* A phone or mobile device designed with a CDMA radio. Unlike GSM, with its removable SIM-based architecture, the identity of the CDMA subscriber is bound to the electronic serial number (ESN) of the phone itself and is typically under direct control and management of the sponsoring carrier. In effect, all CDMA phones are locked to a specific service and carrier, with little option of reuse on a foreign network.

- *Cellular base stations.* Control the transmit/receive portion of the network and are strategically located throughout an area to provide the RF coverage necessary to support the cellular traffic.

- *Network switching center.* Compares to the MSC of the GSM network in performing all management and routing functions for active devices on the network.

- *Network switching gateway.* Provides an interface to the PSTN.

- *Home location register (HLR).* Provides user and device subscriber information for authentication.

2.1.3 Cellular Service Deficiencies, Challenges, and Opportunities

Regardless of the specific technology, mobile communications provided by wide area wireless carriers are limited in coverage. Aside from the geographic coverage limitation, the major deficiency is lack of complete coverage inside buildings (offices, healthcare facilities, malls, and the like). Once you are inside many public buildings, cellular coverage is blocked by RF opaque walls. If your mobility solution depends on cellular services, that mobility functionality may be lost once you go inside. Carriers clearly recognize this problem as a weakness in their offerings. "Fewest dropped calls" is the tagline of one major U.S. consumer carrier. To a certain extent, not providing coverage inside buildings limits the growth of the carrier business where more cellular "minutes" could be used. Addressing this problem means either bringing the "outside in" or the "inside out." In the former scenario, cellular coverage would be extended inside offices, schools, and public buildings. In the latter case, popular "inside" wireless coverage would be linked with the carrier network. Both options are being vigorously pursued.

The deployment of *femtocells* or *picocells* is a concept whereby a carrier base station functional equivalent is deployed inside a building to extend the carrier network coverage. There are no technical barriers with this approach but rather a significant financial and physical deployment logistical problem that needs to be addressed for such a deployment model to be successful. Collaboration with the facility network management team is important because of increased management responsibilities, compounded by WiFi managed services by the facility owners. Additionally, the question "Who pays for the picocells?" needs to be answered because they can be quite expensive. The hosting facility might not receive any direct monetary benefit from the extension of the cellular network coverage. How does this relate to an ROI decision model? More information on this approach is found in the following chapters.

A standard that defined how to bridge an "inside" WiFi connection to a cellular connection was incorporated in the *de facto* 3GPP v6 standard: Universal Mobile Access (UMA). The base concept promoted through this group is that GSM signaling and audio streams are tunneled through a WiFi connection when available. Whether inside a building or outside in full cellular coverage, the handset retains its functional behaviors as supported by the carrier. This approach is termed a *carrier-centric* approach because the call control remains inside the carrier "cloud." The Fixed/Mobile Convergence Alliance[6] (FMCA), a collective of worldwide cellular providers and chip/handset vendors, was formed to promote 3GPP standard and associated FMC solutions.

Whether the carrier coverage is brought "inside" a building or a WiFi transport can be used to extend the wireless cellular coverage, one form of UMC can be achieved. The "outside-in" coverage strategies will be promoted by the carriers wanting to extend their market shares and increase their subscriber ARPU. The "inside-out" coverage strategies will be promoted by PBX, networking systems, and wireless LAN vendors. The success of either approach will be determined in the marketplace over the next six to eight years, and the consumer/enterprise will have multiple solution choices from which to choose. More details describing each approach are found in subsequent chapters.

2.2   Wireless LANs

The initial motivation of the IEEE 802.11 Wireless LAN (WLAN) standard was to provide a simple wireless alternative or extension to hardwired Ethernet. As these efforts predate the advent of wireless VoIP and IP-video application requirements becoming important, the initial WLAN standard lacked provision for both quality of service (QoS) and adequate transport security support for such applications. Early WLAN deployments were implemented with the focus of providing *portability* versus mobility features. For example, a laptop could be used in one room and moved to another, remaining "attached" to the network (mobility). However, retaining a network connection while being "on the move" and support for real-time applications were not the initial focus of the early commercial products.

Despite the lack of key features like QoS and robust security, early innovators saw a market opportunity for providing wireless voice over WLAN and developed solutions that augmented the WLAN deficiencies with proprietary solutions (see Figure 2.3).

---

[6] www.thefmca.com.

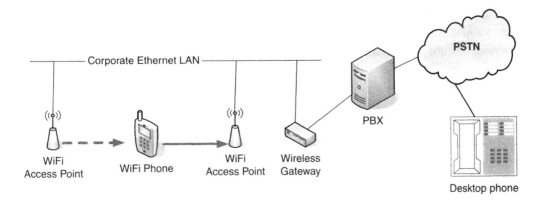

Figure 2.3: Wireless LAN VoIP architecture.

As early as 1998, vendors began promoting 802.11-based wireless phones[7] with proprietary extensions to the 802.11 standard. The SpectraLink Voice Priority (SVP) architecture was heavily promoted and adopted by most WLAN vendors as a QoS solution. Security "holes" were plugged by proprietary extensions from Cisco and Symbol Technologies, but adoption was spotty due to a lack of intervendor interoperability.

In short order, the IEEE 802.11 working group recognized that the standard needed to be amended to address these feature shortfalls. The QoS solution was addressed in the 802.11e amendment, and the security solution was addressed in the 802.11i amendment. Such extensions to an international standard do not happen overnight; it took several years to ratify these new standard elements.[8] Even after these amendments to the standard were ratified, additional product development steps were required. After a vendor enhanced its product to conform to the new standards, there was the matter of validating intervendor interoperability.

Enter the WiFi Alliance (WFA). This industry consortium was formed for the purpose of validating intervendor interoperability for 802.11-based products. The consortium created the term that is often abbreviated to *Wi-Fi* or *WiFi* and is used to describe WLAN implementations.

---

[7] SpectraLink and Symbol Technologies were shipping 802.11b WiFi handsets in 1998.
[8] The 802.11i was ratified in 2004, and the 802.11e was ratified in October 2005.

With the product branding of the WFA, WLAN buyers are assured that products have been validated to conform to specific levels of the current 802.11 standard. As new amendments are ratified and added to the 802.11 standard, products must be recertified to validate conformance with the most recent standards. As the WFA evolved, it saw that the alliance could be more than just a "validation" body and became aggressive to move the WiFi market ahead by launching marketing initiatives on their own. Prior to the ratification of 802.11e, the WFA announced its Wi-Fi Multimedia (WMM) program; prior to the ratification of 802.11i, the WFA announced the Wi-Fi Protected Access (WPA) program. These programs codified specific features of the individual amendments that were not likely to change in the final draft and became the basis of the validation suites. Such programs could move the market ahead because the assurance of functional interoperability was guaranteed by an acknowledged industry body.

Though most WLAN infrastructure vendors quickly implemented support for the standard acceptance criteria, many of the mobile handset vendors lagged behind. This has been primarily because the voice or video market had not matured for this class of applications. However, as the market expands, it has become important that such devices be on a functional par with laptop or infrastructure wireless LAN products in their support of QoS and security functionality. More detailed discussion regarding these functional components is found in the following chapters.

The IEEE 802.11 standard continues to evolve with regard to support of new features and functions. Fast roaming (802.11r), neighborhood reporting (802.11k), and mobile client management (802.11v) are all emerging amendments; such features, when the amendments are ratified, will add robustness to wireless technology, expanding its usefulness.

A number of 802.11 telephony devices are available on the market today. Typically these are offered as solution components in conjunction with WLAN or VoIP product families. Voice quality is often optimized through proprietary mechanisms that minimize the vendor selection options available to the market. The following chapters discuss the key elements necessary to achieve the vendor-independent ecosystem that is required for successful UMC solutions.

2.3   Clarifying Popular UMC Product Terms

A number of terms are used to describe the various UMC solutions emerging on the market. Whereas UMC is the umbrella term for the general category of mobility solutions, FMC has become a general term describing a specific class of implementation approaches to the mobile communications challenge. What follows is a brief description of each of these major product terms to help the reader frame an understanding of the technical portions of this work. All provide some level of handoff between WiFi and cellular networks but may vary in architecture where the call is anchored and in who manages the calling environment.

2.3.1   FMC Definitions

Perhaps the most popular UMC term used in industry publications and by companies promoting mobility solutions is *fixed/mobile convergence* (FMC). The use of the FMC term has become popular for describing many different mobile solutions that might not have the same architecture or feature set.[9] Without some investigation of exactly what any one vendor may provide with its FMC solution, it might be assumed that all FMC solutions are alike; *this is not the case*. Because the FMC label is so predominant in industry publications and product literature, the following section is provided to describe some of the architecture and feature variances that may bear this label.

The FMC term was initially coined to describe the high-level function whereby a single dual-mode wireless device could be made to bridge a call between the traditional PSTN (fixed) telephony network and the cellular (mobile) network. Some early entries in the mobility market attempted to describe the ability to "transfer" a desktop phone call to a cellular phone as a form of FMC. One major telephony provider even made a press release describing its FMC market offering that consisted of its VoIP-SIP (fixed) and WiFi-Only (mobile) support. However, the offering had no support for a dual-mode handset. Another major VoIP provider promoted its FMC solution that was nothing more than a schizophrenic design providing two phone modes that executed independently, one inside the building and one outside. This kind of communication can confuse the interested buyer. Other products would offer a "hybrid" model whereby outbound calls could be made over WiFi and inbound calls over cellular. Yet all carry the FMC label.

The intent of most FMC solutions is to support the capability of performing a call handoff on a single device between the two public or private wireless networks. Optimally, this operation should be designed to be seamless and would appear to the

---

[9] Moving into the future, the FMC term may be replaced by the Mobile Unified Communications term, which has a broader span in describing mobile communication functionality.

user to be converged with no awareness of the underlying active transport network. Unfortunately, the FMC term is overloaded and is not used in a consistent manner to describe products emerging on to the market. Though there is a core of common functionality with all such products, the implementation and usage models often vary, falling into five basic design classes:

2.3.1.1   FMC: Literal Fixed/Mobile Convergence

The original goal of the FMC efforts was to define an architecture by which a call that was serviced by a wireline carrier (circuit-switched fixed) could be handed over to a cellular phone serviced by a mobile carrier (see Figure 2.4). Early FMC offerings were aimed at meeting this base-level definition for FMC via a manual transfer operation, but this model does not completely address the mobility required in today's market.

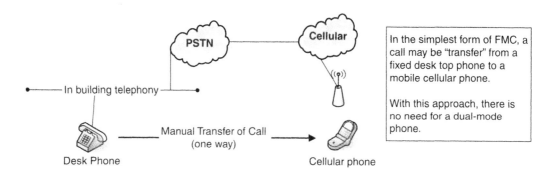

Figure 2.4: Manual fixed-to-mobile transfer.

Complexity was a challenge because it involved collaboration of both wireline and cellular providers for support of the "transfer" and arbitrage of any resulting service fees. This condition was easily met where a wireline provider was also a wireless provider, but few such multiservice providers exist and the call transference was only one way (fixed to mobile).

2.3.1.2   FMC: Find Me/Follow Me

A simpler and more popular approach for providing an FMC-like functionality is offered by many PBX vendors with their "intelligent" find-me/follow-me feature. By extending normal PBX no-answer, busy, or unconditional transfer configuration

options, a cellular phone number may be associated with a PBX extension
(see Figure 2.5). When a call to a PBX extension is processed and the call state
meets certain criteria, the call can be "transferred" to the associated cellular phone. This
model goes halfway toward meeting most mobility requirements, dynamically
redirecting landline calls to a mobile phone.[10]

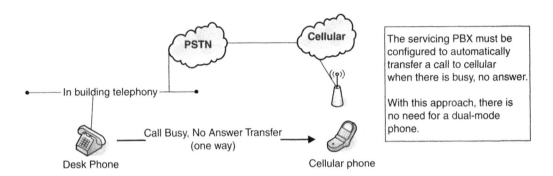

The servicing PBX must be configured to automatically transfer a call to cellular when there is busy, no answer.

With this approach, there is no need for a dual-mode phone.

Figure 2.5: Auto fixed-to-mobile handover.

Capabilities like this are not of much benefit to the average consumer but can be of
great benefit to mobile enterprise associates. However, such offerings do not provide
a fully integrated solution, because outbound calls from the cellular phone are still
handled solely through the cellular network, with no coupling to the enterprise PBX.

2.3.1.3 FMC: Manual Handoff

With the advent of dual-mode phones (WiFi & cellular), a more consistent FMC
architecture could be implemented, one in which a single device could be used that was
attached to the fixed network through a WiFi WLAN infrastructure and to the cellular
network (see Figure 2.6). In such a configuration, it is possible to create client software
that manages transitions between disparate networks.

Not only could an inbound call to a PBX extension be accepted by an *on-premises*
wireless phone (WiFi attached), but that same phone could travel off-premises and have
the call handed over to the cellular network on the same physical handset. Handoffs
crossing wireless domains (WiFi-to-cellular and cellular-to-WiFi) are possible, greatly

---

[10] Avaya supports its Extension to Cellular (EC500) option based on this model.

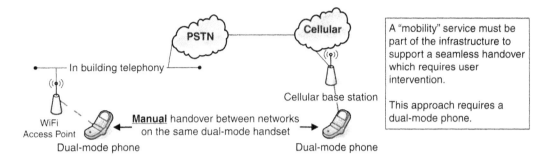

Figure 2.6: Manual handover model.

extending the mobile functionality. The simplest implementation of such a bidirectional model is one that implements a manual handover operation. When the user gets to the edge of the selected transport network, he is alerted and can invoke a procedure that takes 10 to 15 seconds to hand over the call.[11] This FMC option, though affording more mobility, is somewhat clumsy in requiring the user to make a network-aware decision on handover.

2.3.1.4   FMC: Enterprise Seamless Handoff

The ideal FMC design is one that performs seamless, bidirectional handoffs between WiFi and cellular networks, without user intervention. With such an architecture, the call is never interrupted during a handover (see Figure 2.7).

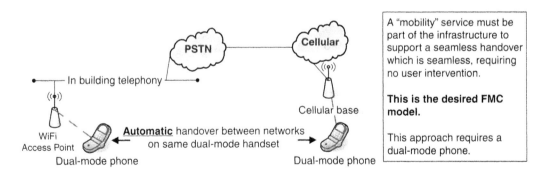

Figure 2.7: Automatic handover model.

[11] An early Avaya one-X dual-mode client was an implementation that used a manual handover design.

Such is the architecture proposed by the 3GPP standards body and by a number of mobility solutions coming on the market. This core feature makes no assumptions as to where the call may be anchored (PBX or cellular) but provides the user with a seamlessness not offered with the other implementation options. Not too surprisingly, this later approach is more complex and has more market and technology barriers to address.

2.3.1.5 FMC: Carrier-Based Seamless Handoff

Carrier-based handoff and UMA and voice call continuity (VCC) provide an automatic handoff between WiFi and cellular networks but differ where the call control is resident. In the case of a UMA solution (see Figure 2.8), the dual-mode phone uses the basic IP transport of a WiFi access point in the same manner that it would use a standard cellular mobile switching center (MSC).

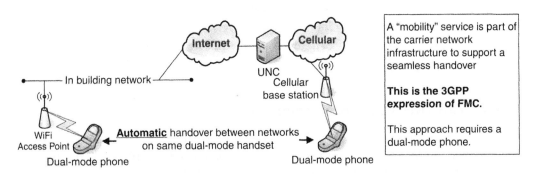

Figure 2.8: Carrier-based FMC handoff.

2.3.2 UMA Architecture

Universal Mobile Access (UMA) is a term used in the 3GPP v6 standard to describe one of the sanctioned FMC solution approaches. The current term was changed from its original *Unlicensed Mobile Access*, which may be somewhat confusing to the casual reader but is now more descriptive of the proposed ubiquitous nature of the application.

Technical details of this solution design are provided in Section 5.4.1 of this book, but this GSM-specific design describes a means by which cellular audio, packet data, and signaling may be tunneled over a WiFi link back into the cellular "cloud" through a *network controller*. In this case, the WiFi radio operates like a cellular radio, delivering the same mobile applications to the user, regardless of the network. Typically, this approach is more of a consumer play than an enterprise or business solution. UMA is used as the technology in dual-mode handset service offers from T-Mobile in the United States; Orange in France, Spain, the United Kingdom, and Poland; Telia Sonera; Rogers; and Cincinnati Bell.

2.3.3   VCC Architecture

A proposed addition to the 3GPP standard set is the Voice Call Continuity (VCC) design. Like UMA, the VCC solution provides a seamless handover between WiFi and cellular with a dual-mode phone, but unlike UMA, VCC specifies that the client be Session Initialization Protocol (SIP) based and not necessarily GSM dependent. One advantage to the VCC approach is that its underlying architecture is potentially protocol compatible with the emerging IP Multimedia Subsystems (IMS). This underlying design feature provides some flexibility for these solutions and a network transport agnostic potential. As currently specified, however, VCC implementations have functional limitations not suffered by a UMA-based dual-mode handset service.

2.3.4   CBC/MMC/eFMC Architecture

Cellular-Broadband Convergence (CBC), Mobile-to-Mobile Convergence (MMC), and Enterprise FMC (eFMC) are terms coined by new vendors offering mobility solutions that are standards based running on dual-mode handsets but are carrier agnostic. The desired seamless handoff between WiFi networks and cellular networks is supported, but no feature or service dependencies are placed on the hosting cellular carrier. GSM networks that are 2G, 2.5G, and 3G are supported, as are CDMA networks. Products of these classes are typically not targeted to the general consumer but rather a business-class user where the mobility service is managed by the enterprise or a third-party hosting service provider.

2.4   Are Customers and Vendors Ready for UMC?

Available mobile technologies have come a long way over the past 20 years. Cellular coverage is assumed in most metropolitan areas, and WLAN technologies have been broadly embraced by businesses and consumer-facing services. Many hotels, airports, and coffee shops now have WiFi "hotspot" service as part of their amenities. Also, more municipalities are deploying public WiFi services as a matter of convenience for the public; in Philadelphia, for example, more than 135 square miles of the city are covered in WiFi, and new city-sponsored WiFi services are being added each month.

Beginning in 2007, several regional pilots for UMC were launched by the more aggressive carriers. T-Mobile and Sprint have both launched FMC services in North America, and T-Mobile and France Telecom launched FMC programs in Europe. However, given the state of the technology and availability of wireless services, there are still limited carrier sponsored commercial services supporting UMC today. Most cellular users are still prevented from making calls inside buildings, and UMC solutions are not broadly available in WLAN services to bridge the network gap and provide a seamless transition across the two wireless domains.

2.5   Analyst Predictions for UMC

As the momentum builds around the topic of mobile communications, industry decision makers and innovators alike look to the analyst community for an indication of the trends that will drive this market. The key submarket components that are important in driving the success of UMC will be:

- Dual-mode handset market

- WiFi market

- Adoption of VoIP (specifically Session Initiation Protocol)

2.5.1   Dual-Mode Handset Growth

A number of respected analysts have been aligned in their projections for phenomenal growth for the dual-mode handset market. Disruptive Analysis projected that a dual-mode SIP phone and UMA market would conservatively be almost 35-40 million handsets per year by 2009 (see Figure 2.9). On the aggressive side, ABI Research anticipates that the dual-mode handset worldwide market might even grow to exceed 100 million phones per year by 2012 (see Figure 2.10). Infonetics has added its perspective that dual-mode WiFi/cellular phone sales will have a compound annual growth rate (CAGR) of almost 31% over the next four to five years and will grow to almost a $3 billion business by 2010.

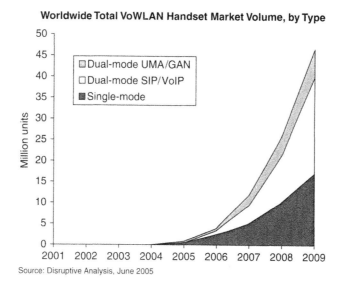

Figure 2.9: Disruptive Analysis dual-mode forecasts.

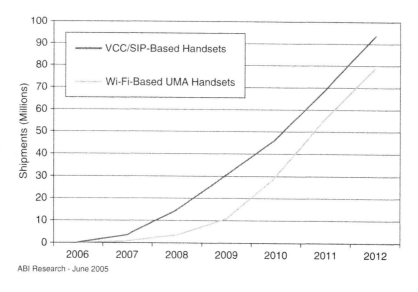

Figure 2.10: ABI Research dual-mode forecasts.

The general consensus of all the major analysts is that dual-mode phones will become a standard offering from phone manufacturers going forward. This has been validated by Nokia's commitment to develop *only* dual-mode phones for future offerings.

2.5.2 WiFi Market Growth

Strong growth is projected for deployment of WiFi technologies in private, public, and commercial business. A Juniper Research report states that the enterprise WLAN switch/mobility controllers (supporting VoIP) market will reach almost $8 billion by 2012. Additionally, In-Stat/MDR reported like continued growth in the worldwide WiFi market at a growth rate of about 14%.

2.5.3 VoIP Market Growth

The fact that all the major worldwide PBX vendors have announced their commitment to VoIP coincidental with the abandonment of TDM makes the success of VoIP solutions a fact, *not an "if," but a "when."* Analysts have echoed that view as analyst Barry Butler of Juniper predicts VoIP revenue for enterprises to rise to $18 billion by 2010 with severe impact on wireline carriers. In-Stat (2006) reported that total IP phone annual shipments will grow from 10 million units in 2006 to 164 million units in 2010.

2.6 The Market Is Ready for UMC

Baseline coverage of cellular networks (outside) and deployment of WiFi networks (inside) are rapidly growing in metropolitan areas. Several major corporations have already voted on the essential need for mobile communications for their employees.[12] To be a worldwide market success, however, other solution components must be commercially available at a competitive price point. Collaboration of multiple vendors and interoperability testing of their products must be achieved to fully realize a UMC "world." Chapter 3 annotates the requirements of each solution element necessary to fully deploy UMC.

---

[12] Both Nokia and Ford have converted a large portion of their employees to cell phone only mode and remove their desk phone.

Unbounded Mobile Communications: What Is It?

In a world where the industrialized nations are rapidly approaching per-capita saturation of cellular phones and the deployment of traditional landlines is rapidly declining, the next generation of mobile communication has stepped onto the stage: unbounded mobile communications, or UMC. Also referred to by other names and acronyms, UMC's desired core function is one that supports a cross-technology integration extending telephony (and data) accessibility beyond the limitations of cellular coverage to include support of other wireless technologies such as WiFi (IEEE 802.11) and WiMAX (IEEE 802.16). The challenge of a cell phone roaming between two technology-compatible carriers (for example, two GSM carriers) has been solved, albeit often accompanied by a high roaming charge. The latest challenge is how to create new and innovative solutions that can span dissimilar wireless technologies—"to go where no cell phone has ever gone before."

3.1 What Problem Does UMC Solve?

Current mobile telephony solutions address the most significant problem for both consumer and enterprise mobile users alike: *accessibility*. In a fast-paced and overscheduled world, individuals find themselves on the move and away from traditional "fixed" telephones, yet with a need to be reached by and to contact others. The ability to communicate at will, without consideration of location, is the prime goal for UMC. A secondary problem that is often part of the solution is the ability to roam seamlessly between disparate networks without dropping the call. This affords the uninterrupted continuity that further insures the all-important accessibility: *anytime and anywhere*.

Do the existing telephony products solve the majority of mass mobile communications market requirements? No. Are there other considerations that need to be addressed? Yes.

3.2   Whose Problem Does UMC Solve?

Being accessible for making and receiving phone calls is convenient, but is this a sufficient basis on which to build a multibillion-dollar business? The market pundits have responded with a resounding yes, and at least two distinct submarkets have emerged:

- Consumer mobile communications

- Business mobile communications

Even within the business mobile communications market, there are distinct product requirements for large enterprise customers and small and medium-sized business (SMB) customers. The product differences at this level are primarily driven by the sophistication of the end user and their skills in deploying and managing highly technical solutions. For the SMB, a "managed" (or "hosted") service is often the one of choice, where they opt for a "hired gun" who is responsible for daily operations of such solutions while enterprises "do it themselves."

3.2.1   Consumer Mobility Requirements

The consumer market has overwhelmingly voted to adopt a wireless communications solution. Today it is almost a rite of passage for a teenager to be given a cell phone, which nicely complements the mobility that they have achieved with a license to drive automobiles. To expand their subscriber base, wireless providers heavily promote "family plans" to make sure all members of a family have cellular phones and can communicate with each other. The competitive features provided by most carriers in today's market are sufficient to meet most consumer's peer-to-peer mobility requirements (see Table 3.1).

Each consumer is treated as an independent agent from a wireless telephony perspective, and all call and feature management is wireless network centric. Competition between providers is fierce, and it is to be expected that new demand will be generated through support of additional mobile features such as ringtones,[1] mobile TV, and MP3 music.

---

[1] Strategy Analytics projected that by 2008, worldwide revenues on ringtone sales would reach US$4 billion.

Table 3.1: Consumer mobility requirements

Mobile requirement	Description
Basic telephony	Ability to establish or receive a phone call to or from any other mobile or PSTN hosted phone
Voicemail	Even when one is mobile, not all calls can be answered, thus a requirement for wireless service provider-sponsored voicemail
Three-way conferencing	Most carrier SLAs include some form of ad hoc conferencing
Text messaging	Ability to send/receive text messages from other mobile or desktop phones; SMS: Short Message Service
Multimedia messaging	With the advent of camera phones, the requirement to transmit digital images to mobile or network destinations became real
Push-to-talk	Recently, carriers have offered support for a simplex, one-to-many walkie-talkie-like functionality
Internet access	Basic Internet services are supported over the Cellular Data Channel (CDC), which requires an additional service agreement option per user

The major limitation on a consumer's use of a cellular phone is still one of coverage, whether geographic, outside normal urban areas, or inside buildings where the cellular signal may not penetrate.

3.2.2 Enterprise Mobility Requirements

Mobile business users require a similar base feature set to that required by mobile consumers but have extended requirements necessary to fulfill their job functions (see Table 3.2). Such a mobility solution may be coupled to a company telephony system and network infrastructure, benefiting the corporate user:

- Single number for business contacts or "single number reach"[2]

- Eliminates the need to publish both office desk number and mobile phone number[3]

---

[2] Most mobile professionals need two mobile numbers: one for the office and one for personal use. Successful UMC solutions will support both use modes.

[3] Only the office number is required on a business card, and the enterprise controls the phone use.

- Users have to check only the corporate voicemail system, eliminating the need for cellular voicemail service

- Accessing the corporate phone directory simplifies the overhead of managing a separate phone list.

Table 3.2: Enterprise mobility requirements

Mobile requirement	Description
Business telephony	Ability to establish/receive a phone call to/from any other mobile or PSTN hosted phone that is coupled with the company telephony solution, displaying corporate Caller ID
PBX integrated telephony	Access to the PSTN or other mobile phones would be routed through the corporate PBX: Call transfer Call Hold Call Mute Call Conference Message Waiting Indication
Voicemail	When used as an extension to a company desk phone, it is desirable to access only the corporate PBX voicemail and not any carrier-supplied voicemail
Text messaging	Ability to send/receive text messages from other mobile or desktop phones; must have an enterprise contact-based directory, be secure, and may be a carrier-independent instant messaging (IM) service
Multimedia messaging	This can be important, but the business urgency of this requirement is low; certain verticals may have a strong requirement for this capability
Push-to-talk	Recently, support of a simplex, one-to-many walkie-talkie-like functionality has become important in many business sectors
Internet access	Basic Internet services are supported over the Cellular Data Channel (CDC), which requires an additional service agreement option per user
Vertical application support	Provides a platform for supporting market-specific applications to maximize the ROI of a UMC solution
Enterprise call control	Control over cost and usage policies dictates that the call control be enterprise centric; this includes support of business-specific security requirements and least-cost routing

From a business perspective, the value-add of a UMC solution goes beyond *accessibility* and results in increased *productivity*.

Upgrading a dual-mode device with extended application features drives up the value-add and allows businesses to better manage their mobile communication dollars as well as address many problem areas that block or inhibit their business success.

3.2.3 UMC Solution Applicability

Each market segment product we've described has a place in meeting the needs of specific market segments. However, because all these solutions use similar terms to describe their products, it is important to understand the differences and concepts so that you can make that "best-of-breed" decision.

3.3 A UMC Agnostic Approach

3.3.1 Client Agnostic

Dual-mode (WiFi plus cellular) handsets are now coming to the market from multiple vendors and may be based on a variety of operating systems: Windows Mobile, Symbian, and Linux. Major cellular providers are shipping UMC-capable devices without support for mobile-to-mobile roaming but have made announcements regarding such support in regional trials. One product realization dynamic that is systemic with implementing UMC services on handsets is that the handset market is fragmented and each vendor has its own software development environment, which results in products coming to market with little guarantee of interoperability. Development of mobile clients for these dual-mode handsets would ideally come from a single independent vendor in conjunction with a prime component vendor (e.g., handset manufacturer, carrier, or WLAN partnership). Choice of a handset OS may be driven by corporate policy or programmability of the system. Such a development environment could provide the market with the broadest selection of handset form factors that was agnostic to client OS.

3.3.2 Network Agnostic

To maximize the mobility desired by today's enterprise, a successful UMC solution will provide a consistent set of client behaviors, regardless of network coverage; seamless roaming between WiFi and cellular networks is table stakes. A network-agnostic design will afford a product that will operate in any one of the popular public and private wireless network technologies: GSM, CDMA, WiFi, and/or WiMAX. The most successful UMC solution will offer customer choices in network configurations that will support multiple handset options that have been designed to support different network technologies. A consideration of an international corporation will also be the requirement of supporting different carriers, because of the country-specific carrier dependencies. In such a case, the UMC vendor of choice would be one that supported all the carrier networks of interest.

3.3.3   PBX Agnostic

Seamless roaming across diverse networks affords limited value to mobile enterprise associates unless it is also coupled with a critical enterprise application. Telephony is the lifeblood of enterprise, and being able to interconnect with a corporate telephone communications system while mobile is important. The majority of enterprises have chosen to host their own PBX or iPBX, and a successful UMC solution provides an integration path with a broad set of such products. This poses integration strategy challenges because the installed base of PBX systems is still mostly based on Time Domain Multiplex (TDM), and such configurations require an additional "gateway" product to bridge between the PBX and the mobility appliance controller. Even with iPBX systems that are SIP based, there is often an integration problem where the mobility appliance and the iPBX were implemented with different interpretations of the SIP standard. Such hurdles are not insurmountable and will be addressed by the successful UMC solutions.

3.4   UMC Handover Logic

Considering the design assumptions and goals we've discussed, a tailored mobility solution can be described and consists of two major components: (1) a network mobility appliance and (2) a mobile client. These two elements work collaboratively to provide a seamless experience that is network agnostic. It is the UMC client that is instrumental in driving the network roam decisions, and hybrid logic is implemented across OSI layers to accomplish this task (see Figure 3.1).

The handset WiFi drivers (layer 2) are responsible for managing the smooth Access-Point-to-Access-Point roaming. To provide the best voice quality, such roams must be in the sub-50 millisecond range. Additionally, it is the responsibility of this layer to

WiFi⇔Cellular Handover Decision Architecture

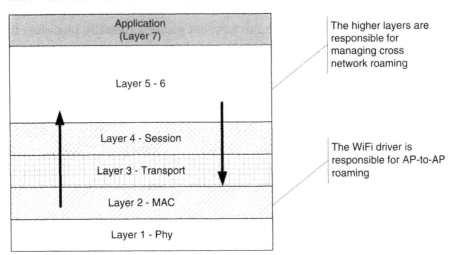

Figure 3.1: Handover logic hierarchy.

enforce the appropriate wireless 802.11i security policy: WEP, WPA, WPA-PSK, WPA2, or other. As the IEEE 802.11 standards exist, there are still "holes" in the service fabric that need to be addressed to define fast roaming (802.11r) and traffic load balancing (802.11k and 802.11v) within the WiFi space. Enhancements in the layer 2 support features for voice applications will continue to evolve over the next few years but may come to market asynchronously on a vendor-by-vendor basis.

To provide the seamlessness essential for mobility, logic at the higher levels (layers 5 and 6) monitors the status of the viability and health of the two networks (WiFi and cellular) and decides which network will provide the most reliable connection and best voice quality.

Establishing and maintaining a phone call across two dissimilar networks is a balancing act. For calls established over WiFi, the UMC solutions would most likely implement Session Initiation Protocol (SIP, IETF RFC 3261). This is the market-dominant VoIP standard and affords great flexibility for integration and functionality options in today's market. Whether a call is set up while in the enterprise corporate LAN or remotely through a WiFi hotspot connection, such IP-based calls may be augmented by certain UMC secure extensions to manage functions such as registration/authentication, presence, and messaging.

For calls utilizing the cellular network, UMC products often take advantage of a carrier's standard feature set for an enterprise-centric product. Certain UMC signaling may be transmitted over packet-data services, but audio is typically processed through the standard bearer channels. Use of these infrastructure components guarantees the best voice quality while providing the control elements needed for traversing the networks.

Carrier-centric UMA products have a fairly straightforward architecture with regard to the way the WiFi networks are integrated into the overall carrier network. Since the WiFi services provide a tunneling transport, all signaling and audio media transmission merely emulates what happens in a "native" cellular network. VCC carrier products support SIP but would also follow a "converged" data stream model of commingling signaling and audio straight into and from the carrier network.

One optional design element in supporting enterprise-centric UMC seamless roaming is a "hinged" architecture approach.[4] Traditionally, pure VoIP implementations do not couple call signaling with audio streams, but to better manage seamless roaming and to provide value-added functionality, traffic for each call may be routed through the mobility appliance (see Figure 3.2). In this case, the UMC client collaborates with the

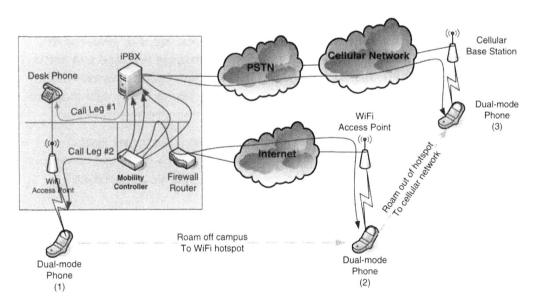

Figure 3.2: Hinged call architecture.

[4] May also be called *tromboned.*

UMC mobility appliance to monitor the overall voice and link quality to ensure optimal voice quality and reliability; should either link degrade below predefined threshold levels, action is taken to roam that "leg" to an alternate network (WiFi or cellular), without user intervention, and the call will be sustained until either user terminates the call.

Figure 3.2 is an example of a "hinged" call that is the basis for most enterprise-centric UMC products. Unlike a carrier UMC implementation, the control point of an enterprise-centric solution resides within the hosting enterprise network, and all traffic between a mobile handset and a peer landline phone traverses the network through the mobile appliance. This is a unique requirement to support the ability to roam between WiFi and cellular networks.

In the example, a mobile phone (1) initiates a call with a desktop phone and then walks out of range of the corporate WiFi into the coverage of a public hotspot (2). The call path is "hinged" in that there are two independent audio legs that are established for this one phone call: one from the desktop to the mobile appliance and one from the mobile appliance to the mobile handset. The mobile user moves out of range of the public WiFi (3) and establishes a standard cellular call back to the mobile appliance. As the mobile user roams from the in-house WiFi to a published WiFi source, the "mobile" leg is reestablished through the new connecting network services and the original path is turned down. The "leg" between the desktop phone and the mobile appliance is sustained, regardless of how much the mobile handset roams across networks. Through such a "hinged" architecture, it is possible to decouple two parties in a call, allowing for unrestricted mobility on the part of each party.

- *Handover from WiFi to cellular.* When a call is established in WiFi coverage, the call is monitored as to Received Signal Strength Indicator (RSSI), error rates, congestion indicators, and jitter metrics. If these parameters are found outside defined ranges/thresholds, the client will invoke a "roam" to the cellular network. On receipt of the negotiated inbound cellular call, the WiFi leg is terminated and the call automatically switches to the cellular audio stream without user intervention.

- *Handover from cellular to WiFi.* If a call was established in the cellular network, the client will periodically query for an authorized WiFi network. Preferentially, if WiFi is located, the client will communicate with the mobility appliance to initiate a roam from cellular. Insertion into the hosting WiFi network may require conforming to WiFi security policies (WEP, WPA, and so on) in addition to other enterprise security policies. The client is responsible for navigating these hurdles and will switch audio streams once a SIP call is established.

Other UMC enterprise products introduced to the market may choose to segregate signaling from media. With these implementations, some performance advantages can be achieved by minimizing bottlenecks through the mobility appliance but will suffer loss of application monitoring, control, and extended feature support.

3.5   UMC Alternatives

The solutions described in the previous sections utilize dual-mode handsets and provide virtually unlimited extended coverage for a mobile user anywhere there is WiFi and anywhere there is carrier service. An alternative being promoted by the wireless carriers employs a concept called *femtocells* or *picocells* to extend the cellular network coverage into commercial or private residents. In principle, a *femtocell* is a low-cost, low-power wireless carrier base station that is installed inside a hosting building with an IP/Internet link back into the carrier cloud (see Figure 3.3). Because these "cells" emulate a standard cellular base station, a standard cell phone will continue to operate in the same manner that it does when outside in full network coverage.

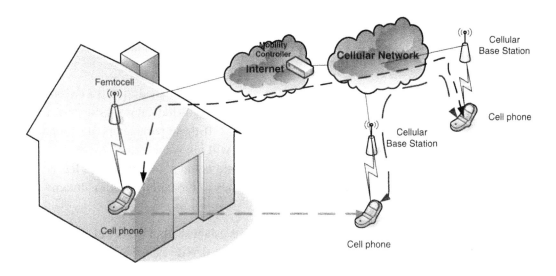

Figure 3.3: Femtocell mobile solution.

Some industry analysts[5] predict that over 100 million cellular users will take advantage of worldwide femtocell deployments by 2011. There are also a number of femto/picocell commercial offerings that are currently on the market, most of them targeting the home, where a single femtocell will suffice for extended coverage.[6] Support for 911 is supported as any cellular phone by triangulation from nearby base stations. This technology will extend the carrier network and minimize subscriber "churn" (a big problem); the overall success of this approach will be determined by cost and individual consumer needs. The problems encountered in attempting to deploy a femtocell infrastructure within an enterprise, however, are immense and potentially prohibitively expensive.

3.6   The Mobile Enterprise

As the industrialized work becomes more mobile, the demand for mobile communication increases. It is clear that the early adopters of this mobility will be the enterprises because they can make a strong ROI case. Lower CAPEX and OPEX drive such decisions, but increased associate productivity and customer satisfaction are also realized when UMC solutions are deployed. Being competitive in the 21st century requires aggressive deployment of tailored mobility solutions.

---

[5] ABI Research, 2007 report.
[6] Sprint's new Airave is a femto cell connected to the owner's home DSL line.

CHAPTER 4

UMC: An Overview of Technology Requirements and Considerations

To achieve a goal of delivering truly "unbounded" mobile communications solutions, a number of technologies must be aligned. Delivering a seamless and secure communication path from handset, across multiple networks, to handset that is agnostic to network technology or geography requires significant enhancements and improvements to today's technology offerings and equal evolutionary changes in the ability to deliver such solutions to the customer through existing market channels.

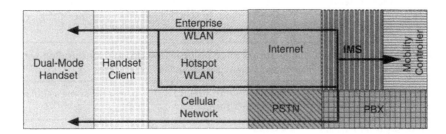

Figure 4.1: UMC technology set.

Figure 4.1 outlines the multiple paths that signaling and audio must traverse to deliver a true end-to-end UMC solution. Entering the first decade of the 21st century, no single vendor provides point products that span all these technologies. Certain major industry players may provide an 80% solution, but there are dependencies on other solution components that must be supplied and supported by third-party vendor products and services. All these components must be "knitted" together into a seamless solution before unquestioned adoption by the market will occur. Each sub-element of the UMC solution must meet certain baseline requirements to contribute to its portion of the overall solution. These requirements and considerations are outlined in this chapter.

4.1   Mobile Handset Requirement

WiFi/Cellular handsets began appearing on the market around 2003 or 2004, with the WiFi design primarily focused on providing wireless access to the Internet. Little or no design thought had been given for these early offerings in terms of use of the WiFi service for voice (or real-time) traffic, and handset cost was high ($600–800 range). Beyond an acceptable cost point, a number of additional characteristics must be supported by the ideal UMC handset (see Table 4.1).

Table 4.1: Ideal UMC handset features

Feature	Remarks
Popular OS	Appropriate handset operating system (Symbian, Linux, MSC Windows Mobile, RIM Blackberry, Qualcomm BREW, or proprietary)
Carrier-independent offering (unlocked)	Traditionally, cellular phones have all been sold through the carrier channels, which limit the customer's choice of service and device
Acceptable market price point	Optimal price point below US$400 or bundled with aggressively priced SLA
Dual-mode Radio Design	Supporting an 801.11 (a or b/g) and some cellular service (GSM or CDMA)
Voice-optimized WiFi Layer-2 services	Secure, pre-emptive AP-to-AP roaming in under 400 milliseconds
Flexible audio output design	Dynamic programmable routing of audio between front speaker and back speakerphone
Acceptable battery life	3+ hours of talk time with 24+ hours of standby time
Acceptable form factor	Candy bar or PDA with proper durability requirements, including support of acceptable keyboard services (phone or PDA class design)
Security	Both device and network access security are critical buy decision elements

A number of factors will come into play as any dual-mode handset comes to market. General market acceptance criteria will be based on price and wireless support. Without an aggressive price point somewhere at or below the price of a desktop phone, dual-mode phone market success will be limited. Most cellular phones have a market life of between 12 and 18 months, which further complicates any broad acceptance and rollout of UMC solutions. Similarly, selection of the desired cellular carrier service may be regional and dictate a fixed buy requirement. In the current market, there are many more GSM dual-mode devices available than CDMA, a fact that limits regional success in North America.

4.1.1   OS Considerations

Multiple dual-mode phones are on the market based on different operating systems. Linux, Symbian, the proprietary Research in Motion (RIM) Blackberry, Google Android, and Windows Mobile are the most popular. All the devices come with a "native" dialer that has been carrier certified, but these do not support any UMC (cross-network roaming) functionality and are strictly cellular service products. Other products may now offer a SIP softphone, but it is not integrated into the "native" dialer and is severely limited in functionality. With the existing crop of commercially available dual-mode phones, any UMC functionality has to be delivered as an adjunct application. The exceptions are UMC phones offered by specific wireless carriers that are "locked" to their service (e.g., T-Mobile's @Home program).

Without direct collaboration with the handset manufacturer or OS provider (Microsoft or Nokia), supplanting the native dialer with a UMC solution is not without its problems. For both Microsoft Windows Mobile-based and Symbian-based devices, the native dialer functionality is built into the OS structure and cannot be treated as a user application. For that reason, any UMC application must be designed to work collaboratively with the native dialer. Even in the best scenarios, there are behaviors that are not optimum. For example, management of reporting of the cellular signal is managed by the native dialer on Windows Mobile devices. A UMC application will "push" this dialer into the background, but while in this state, the strength of the cellular signal will be inaccurate. Other complications may arise from contention for keyboard resources on the device where the native dialer locks a keyboard resource that may be needed by the UMC application. There are workarounds for these kinds of situations, but none of them are clean and they will not provide the desired user experience.

An optimistic view of these classes of products is that platforms such as Windows Mobile and Nokia/Symbian are more programmable than other OS types, which opens opportunities for independent software vendors (ISVs) to create new applications that build on the basic value-add of the handset. UMC solutions are such a class of applications that are emerging onto the market, which means that early adopters will be investigating Windows Mobile and Nokia/Symbian solutions more than other, more proprietary offerings.

4.1.2   Carrier-Independent Considerations

Management of the handset sales and support channel has mostly been through the wireless carrier. Since the phones only worked on a wireless network, there was no motivation to build a different delivery channel. This was specifically true for CDMA networks where it was the carrier that negotiated with the manufacturer and branded the phone for its network. There was no personality module for these phones,[1] and the carrier could even dictate certain unique features that were supported by its brand, as opposed to a competing carrier's offering. Sourcing CDMA dual-mode phones for a UMC deployment will most likely come directly through the carrier. Depending on the carrier's stance regarding UMC/FMC applications, there may be functional limitations placed on that model phone.

The utilization of GSM phone is a bit more flexible with regard to bonding with the carrier. By design, each GSM phone has a Subscriber Identification Module (SIM) that is a miniature smartcard containing the carrier subscriber information and other pertinent user data. The phone is identified by its International Mobile Equipment Identity (IMEI) number, analogous to a MAC address of an Ethernet port. The subscribers' profile and phone SLA information are written to the SIM, which can be installed on any valid GSM phone. When this configuration is registered with the network, both IEMI and subscriber information are forwarded through the network for validation. By decoupling the phone (IEMI) from the subscriber (SIM), any user may use virtually any GSM phone by simply installing his or her SIM.

Even with GSM handsets, some planning is necessary in selection of the handset because some phones may be "locked" to the originating carrier. It is also possible that the SIMs may be vendor locked. This would mean that attempting to install a T-Mobile SIM might not work on an AT&T dual-mode GSM phone. If possible, procurement of "unlocked" phones will provide the maximum purchasing flexibility to the end customer.

---

[1] The Electronic Serial Number (ESN) is burned into the phone when it is shipped from the carrier center.

4.1.3   WiFi Considerations

Most commercially available dual-mode phones support the IEEE 802.11b/g standard (2.4 GHz up to 54 Mbps data rate) and not the 802.11a (5.2 GHz up to 54 Mbps) or new 802.11n (up to 300 Mbps) standard. This is primarily due to the availability of low-power chip sets for the b/g standard that integrate well into a small footprint of a handset. Support for the 802.11a or 802.11n standards, though desirable, is not typically found in mobile devices due to a high-power demand resulting in shorter battery life and more complex radio/antenna design.[2]

The data rates of the 802.11b and 802.11g WiFi systems are more than sufficient to support wireless VoIP traffic loads; even a relatively slow 1 Mbps link can support four to five phone calls per individual access point. Modern WiFi infrastructure products can simultaneously support 802.11a and 802.11/b/g devices in a single location, which simplifies deployment and allows desktops and other 802.11a devices to operate over the same infrastructure along with 802.11/b/g devices with virtual impunity.

Many dual-mode devices introduced to the market, however, were found to be deficient in a number of areas when it came to the robustness of the WiFi services. The following is a generalized list of shortcomings of some early commercial products:

- No fast AP-to-AP roaming; needs pre-scan AP environment
- Immature roaming logic based solely on RSSI strength and not other metrics such as transmit and receive error rates or QoS availability
- Limited rate scaling ability; may be fixed data rate
- No voice optimization for security
- Poor battery management design
- No layer 1 optimization; abbreviated collision back-off

As these devices evolve and mature, the overall quality of the handset WiFi services will improve.

Probably the most critical factor to be considered by an enterprise in deploying a WiFi/ VoIP solution is security. Most likely, the selected handsets will be sourced from

---

[2] Motorola Wireless Enterprise Division offers an 802,11/abg mobile handset solution family.

a different vendor than the WLAN infrastructure manufacture. It is important to align their security options to make sure there is a desirable and compatible support option between the two products. Not all handsets will support all security options of a WLAN switch or access point. A key selection criterion will be that both WLAN infrastructure and handset radios support the highest possible level of security that satisfies the end-user security policies.

For the party responsible for installing the WLAN products, it will be important to measure and avoid any building-interfering RF sources. These can be from microwave devices, cordless phones, or nearby WiFi networks with overlapping channels. To ensure the best UMC experience in a WiFi environment, it is important that the target ecosystem be characterized as to the coverage and noncoverage areas within the perimeter of the facility. Reflection off walls and floors can cause "nulls" to be formed where it may appear there *should* be good coverage but, in fact, there is no coverage at all.

At some point, WiMAX will be considered as an alternative to WiFi for selection of dual-mode devices. Such shifts in underlying wireless technologies shouldn't affect the viability of a well-designed UMC solution.

More detailed information for optimizing a WiFi infrastructure for voice is contained in Section 6.2.

4.1.4   Battery Life Considerations

As of 1Q08, third and fourth generations of dual-mode phones have come to the market. To make a UMC handset viable, it will need to support something on the order of three to four hours of talk time and 24–48 hours of standby time. Current cellular-only phones have been finely tuned with regard to battery life and can realize four to eight days of standby time on a single charge. Integrating an application that interfaces with both WiFi and cellular networks places higher demand on battery capacity, and traditional batteries used for cell phones won't have sufficient capacity to provide the desired standby time. The most successful UMC handsets will have an optimized power management system to provide an acceptable talk/standby time profile. All handset vendors should report this level of data on their datasheets and as part of a UMC purchase process. An intermediate solution to battery life problems is to purchase an extended-life battery that most vendors offer as an accessory.

4.1.5   Audio Routing Considerations

Most smartphone and PDA-class devices are designed with a front speaker/microphone pair (front) and a rear speakerphone. From the "factory" many of the dual-mode handsets auto-route any audio associated with the WiFi network to the rear speakerphone because the design assumes applications such as audio from Web applications, MP3 players, and the like. Early UMC developers found that when they tested their applications, a GSM call had its audio routed properly to the front speaker. VoIP calls that were using the WiFi network for audio transport found that the audio was auto-routed to the rear speakerphone. This caused a usability problem in that the source of the sound would change depending on the servicing network. A seamless mobile telephony application must have control over the speaker resource independent of the current wireless transport network. Because this auto-routing was such a low-level design element, ISVs developing UMC applications were frustrated because they had no audio-control API. The lack of such services was due to the infancy of the dual-mode industry that was dominated by the wireless carriers, which didn't see this as a requirement. Beginning in late 2007, handset manufacturers began to work with ISVs and began releasing UMC versions that could properly control the audio output. When investigating a UMC solution, be sure you understand the capabilities of the selected handset with respect to audio-routing.

4.1.6   Form-Factor Considerations

UMC devices are available in one of two basic form factors:

- *"Candy" bar.* A linear form factor that is modeled to meet telephone functionality with minimal data capabilities. Typically, these devices have a "phone" keyboard based around a 3 × 4 numeric key array and optimized for entering numbers. Entering alpha characters on these phones occurs via complex multikeystroke sequences.

- *PDA/terminal form factor.* This form factor is designed more to meet the requirements of data applications and will have a QWERTY keyboard with a touch screen for viewing emails and documents.

The distinction between these device types is blurring as candy-bar phones are designed with sliding QWERTY keyboards or smartphones with larger screens, but no one format fills the full scope of form-factor requirements. Larger enterprises have a need for both form factors, which sets a high requirement benchmark for UMC providers.

UMC dual-mode devices have evolved from the consumer market and are imprinted with the durability requirements for that market. Handset durability becomes a factor in selecting a UMC device in many enterprise locations. The traditional road warrior may be happy with consumer-grade durability, but mobile facility and security users may have a higher durability requirement.

4.1.7 Security Considerations

Security considerations go beyond what is required to protect the WiFi domain, which can be addressed by enabling the appropriate security support: layer 2 (WEP or WPA2). Roam performance optimization may be achieved where WPA2 "preauthentication" can be implemented. However, security must be considered from a total end-to-end perspective, which includes secured signaling and audio traffic (application-specific security) or link security provided through a corporate-sanctioned virtual private network (VPN). Selection of the appropriate UMC handset device should be made after a total assessment of the security options.

Physical security of the mobile device and resident data is another security challenge. If the device is stolen, how do you protect the data? Was it encrypted and password protected? Can you remotely "kill" or "wipe" the device if the unit is stolen?

4.2 Wireless LAN (WLAN) Requirement

Initially, 802.11 products were viewed as a simple wireless extension of the corporate Ethernet LAN. Because VoIP had not become a driving application at that time, there was no implicit consideration in the standard for how to support real-time applications such as voice or video; thus there was no QoS and only limited security incorporated as part of the original standard. Demand for a wireless network product was real and drove the 802.11 standards body to extend the standards functionality to meet the market requirements. Over the past eight years, a number of extended 802.11 substandards have been defined, ratified, and incorporated into commercial products (see Table 4.2). From the resulting "alphabet soup" of alpha descriptors for these emerging standards, a strong core set of products is now available on the market.

"Speeds and feeds" often drive the networking market, and the WLAN market is no exception. The initial 802.11b standard supported up to a maximum 11 Mbps data rate, but this was quickly deemed insufficient by much of the market, and demand for higher data rates drove the standards to new levels. In the 2.4 GHz range, the 802.11g standard

Table 4.2: IEEE 802.11 RF product classes

IEEE 802.11 standard	RF frequency	Channels	Data rates
b	2.4 GHz	14[3]	1, 2, 5.5, 11 Mbps
g	2.4 GHz	14	1, 2, 6, 9, 12, 18, 24, 36, 48, 54 Mbps
a	5.2 GHz	22	6, 9, 12, 18, 24, 36, 48, 54 Mbps
n	2.4 GHz 5.2 GHz	22	Up to 300 Mbps

significantly extended the data rate up to 54 Mbps, which quickly became the corporate WLAN standard of choice. For backward compatibility, most 802.11g devices, both AP and mobile unit, also support the 802.11b standard. In this manner, any early wireless investment was preserved with the deployment of the newer standard. Low-power chipsets made it possible to develop 802.11b/g WiFi handsets, which now dominate the market.

To offer a WLAN solution in a less crowded frequency, the IEEE created the 802.11a standard, which uses the 5 GHz band and newer encoding methods. Unlike the 2.4 GHz standards, the 802.11a standard offers more noninterfering channels at the same 802.11g data rates. Though these are appealing attributes, the adoption of the 802.11a standard has not met optimistic expectations. There are only a few handsets on the market that support 802.11a; most of these also support 802.11b/g.

The demand for speed, however, has not been satiated. Thus, the IEEE went to work defining an even higher data rate standard that would provide rates greater than 500 Mbps. This task group was tagged as 802.11n. The 802.11n standard has received so much market attention and interest that manufacturers are now releasing *prestandard* versions of this wireless technology, with the blessing of the WFA. These early products, however, are not in a handset or PDA form factor, and the jury is out as to whether any handset or PDA form factor will be able to support 802.11n because of its power requirements and extended radio/antenna designs.

---

[3] To minimize interchannel interference, national regulations have been imposed as to which channels may be used. In North America, only three of the channels (1, 6, and 11) are authorized for use.

4.2.1   WLAN Security Considerations

With all the 802.11 (bgan) standards, support of robust security methods is implicit. The initial RC4-based Wired Equivalent Privacy (WEP) was found to be woefully inadequate, so WiFi Protected Access (WPA) was created. The later security option has several suboptions that allow the end user to tailor deployed security to better conform to an organization's overall corporate security strategy. This book doesn't go into the specifics of these security options, but it makes the observations that having more rigorous security protection enabled has a corresponding negative performance impact on real-time applications such as VoIP. This is particularly true with respect to handsets and PDA (circa 2007), where the CPU processing power is limited (200 MHz and below), and adding the burden of host encryption onto the task of processing voice traffic severely impacts the observed voice quality.

4.2.2   RF Coverage Considerations

Beyond simple installation of WiFi access points, an important consideration to ensure a good voice quality experience is to make sure there is adequate overlapping coverage between access points. The connectivity to the network required by a voice application is more rigorous than for a standard data application. If you are browsing the Internet, walking down a hall in the office, and roaming between two access points, the fact that it takes 500 to 1000 milliseconds to complete the call is not a major problem. For a voice application, however, such breaks in network connectivity can severely degrade voice quality and can possibly cause the call to drop.

In considering deploying a UMC solution in an enterprise, it is vital that a site survey be done for voice and not just data. Such infrastructures are often slightly more expensive because of a requirement for more access points, but the resulting improved guarantee for good WiFi voice is well worth the investment. There is an added benefit for data or "portable" users in terms of increased wireless bandwidth—a win-win for all mobile users.

4.2.3   WLAN/Ethernet Topology Integration Considerations

A wireless LAN rarely stands alone and is typically connected to either the Internet (via a router) or a corporate Ethernet LAN. In either case, the wireless components should be taken into consideration in planning the whole network topology configuration. Many network administrators will choose to separate voice and data traffic on the LAN through use of Virtual LAN (VLAN) functionality, which results in networks that are easier to manage from a bandwidth, QoS, and security perspective. Most commercial WLAN products support the concept of multiple ESSIDs per access point, which can then be mapped to a specific Ethernet VLAN ID defined for the hardwired LAN. Such an approach also aids in managing broadcast and multicast traffic, which can have a negative effect on a WLAN's ability to support concurrent voice traffic.

4.2.4   Standards and Regulatory Considerations

It is important to consider products that have received the WiFi Certification from the WiFi Alliance (WFA), which guarantees a known level of conformance to standards and intervendor interoperability. The status of the WFA certification, however, needs to be reviewed with new purchases, since new IEEE 802.11 standards are being ratified and these products must be recertified to conform to the latest set of standards.

Final buy considerations may be affected by national or regional regulations that impact sales of WiFi products. Government and national interest in control of WiFi products has become a factor that can greatly affect the worldwide market. Specifically, a current action considered by the Australia Commonwealth Scientific and Industrial Research Organization (CSIRO) is to sue WLAN manufacturers over the release of pre-802.11n equipment because of an Australian patent infringement. The suit has not been actively pursued, but if it were, it could have dramatic effects on the availability and price of 802.11n products.

In 2003, the People's Republic of China (PRC) announced its specification and declaration of Wired Authentication and Privacy Infrastructure (WAPI) as a Chinese-specific national WiFi security specification. This was done without contribution from IEEE or any other international standards body, and the announcement stated that it was China's intent to force all WiFi products sold in the PRC to conform to this standard and be implemented in silicon. Licenses for WAPI would be under the control of several selected Chinese corporations, and all international manufacturers would have to collaborate with these state-sanctioned vendors to import products into the PRC. World governments, the IEEE, and the wireless industry protested this restrictive country-specific security regulation and sent emissaries to negotiate with the Chinese government. After a series of intense negotiations, the PRC backed off its position of enforcing WAPI on all imported WiFi products but continues to promote its use for education and governmental uses.

4.3   Voice-Optimized Ethernet Considerations

Ensuring that the WLAN component has been optimized for a VoIP application is only one consideration in deploying a UMC solution; the underlying Ethernet network must also be provisioned to support VoIP applications and should include:

- *Layer 3 QoS.* Enables the minimize delay (0x10) Type-of-Service (TOS) processing at routers.

- *Layer 2 QoS.* Enables support of IEEE 802.1p/Q standard for voice traffic.

Both WLAN and Ethernet networks must be configured to support VoIP or there will be little assurance of optimized voice quality. Some IT managers will also want to consider segregating voice traffic from data traffic by partitioning the network into multiple VLANs. This can be effective in further guaranteeing good voice quality but also has the effect of assuming that *all* traffic on a voice VLAN is VoIP, which might not be true.

Off the corporate LAN, network considerations for voice optimization are limited. Accessing the corporate LAN through a remote ISP Internet service requires the subscriber's understanding of the type of QoS implemented by this ISP that will affect voice quality. Additionally, connection types such as Digital Subscriber Line (DSL) can be problematic with regard to support of VoIP traffic. By design, DSL services are asymmetrical; that is, the downstream (from CO-to-subscriber) bandwidth is much larger than the upstream bandwidth. This is because DSL was designed to support Web-based applications for which the downstream traffic was much greater than upstream traffic. When you attempt to run a voice application over DSL, you could experience scenarios in which voice quality on the CO-to-subscriber segment will be good but the CO-to-subscriber segment will be poor.

4.4   Wide Area Wireless Considerations

UMC solutions depend on user access to commercial wide area wireless services to fulfill the promise of being "unbounded." In most urban areas of developed counties, some form of wide area wireless service is available, and any user may subscribe by purchasing a service-level agreement (SLA) from a wireless provider. Typically, these contracts specify service duration of two years with automatic renewal and a monthly base charge for some minute allowance for voice, messaging, and graphic/file services.

Table 4.3: Wireless Carrier Packet Data History

WWAN technology	Generation #1	Generation #2	Generation #3
GSM	GPRS[4] (up to 80 kbps)	EDGE[5] (up to 240 kbps)	UMTS[6] (14 Mbps)
CDMA	1XRTT (144 kbps)	EV-DO (2.4 Mbps for rev 0 and 3.1 Mbps for rev A)	HSPA[7] (14 Mbps)

In addition to a standard SLA, many UMC solutions also require inclusion of a "packet data" or Cell Data Connection (CDC) service as part of the SLA. CDC is often employed for configuration management, signaling, and registration between the mobile device and the management point of presence. It is with CDC that developers have a challenge. There are vast regional and international differences in how tariffs are applied to IP packet services. In North America, some "all you can eat" plans are available, whereas Europe has a relatively expensive fixed kilobyte/month charge for such services. Transferring a small video clip over CDC can result in a frighteningly large charge. Therefore, one critical consideration that is a must in evaluating UMC solutions is their efficiency in terms of use of CDC resources.

Not all generations of CDC were created equal, at least from the perspective of supporting voice applications. Early data packet service options from the carriers had insufficient data rates to support a real-time application such as voice.

The fact that the CDC option allows a mobile phone send/receive IP data packets means that it is also technically possible to execute a VoIP application over this link. Wireless carriers are very concerned about this possibility because it can cause congestion in their networks without a corresponding positive revenue impact. For this reason, wireless carriers have put pressure on dual-mode handset vendors to throttle or eliminate the WiFi capabilities on these devices[8] or specifically block Real Time

---

[4] GPRS: General Packet Radio Services.

[5] EDGE: Enhanced Data rates for GSM Evolution.

[6] UMTS: Universal Mobile Telecommunications System.

[7] HSPA: High Speed Packet Access; has associated High Speed Downlink Packet Access (HSDPA) and High Speed Uplink Packet Access (HSUPA).

[8] The European version of the Nokia E61 was a dual-mode phone, whereas the North American equivalent, the Nokia E62, was an identical design without the WiFi radio. Pressure from carriers blocked the marketing of this configuration.

Protocol (RTP) VoIP traffic on their network. Some wireless carriers have implemented VoIP "firewalls" using deep-packet inspection to block attempts to run a real-time application over the packet data service. Well-designed UMC applications will not employ VoIP over CDC for this very reason.

Of equal concern with the CDC usage charges is potential carrier-roaming charges for the many mobile road warriors. Unrestricted roaming between carriers can result in surprisingly high charges, particularly when traveling internationally. Since there is no *global* cellular carrier, international travel is fraught with extremely high service provider roaming fees, of $1.29/minute and higher! Globetrotters have resorted to managing multiple SIMs (for the country of the "day") to manage the costs. Such a scenario, however, can quickly become a nightmare to the international traveler who must swap SIMs when crossing country borders. This option has the result of changing your cell phone number to a country-specific number, making it more difficult for people to reach you when traveling (e.g., "When I'm in France, call *xxx-xxxx-xxx*, and when I'm in Italy, call *yyy-yyyyyy-yyy*"). Some carriers have offered more global-friendly service agreements with reduced international roaming charges, but these fees can still have a negative impact on usage costs.

A successful UMC product will allow the user (or administrator) to specify a "black list" of carriers into which roam attempts should be blocked, a network architecture that provides a "local" attachment point which would bypass any roaming requirements, or preferentially seeking WiFi services for those *free* calls. Use of such a feature is a balance between cost and functionality and is a customer-specific value judgment.

4.5   VoIP Requirement

Traditional circuit-switched telephony has been embedded into our societal fabric for over 100 years. Alexander Graham Bell (and others) had no idea of the social, business, and governmental impact that would result from his telephone patent. The ability to speak to someone in another building, another part of town, or across the world is no longer a luxury but a necessity for the general public. However, the disruptive concept of transmitting a packetized voice over a local area network or the Internet was a radical idea that came into its own in the last part of the 20th century and the beginning of the 21st century. As implemented, this new technology is based on Internet Protocol (IP) and thus was labeled Voice over IP (VoIP; see Figure 4.2). Like all disruptive technologies (the automobile, the telegraph, the printing press, the electric car, and so on), VoIP had its dependencies, weakness, and challenges that needed to be addressed before being embraced by the general public.

Access to the Internet has become such an integral part of life in industrialized countries that the demand continues to drive the ever-expanding worldwide communication network. This infrastructure now becomes the "superhighway" on which real-time applications such as VoIP can operate. Seeing the immense opportunity for VoIP products, the telecommunications industry responded quickly with VoIP product announcements from traditional PBX vendors and startup companies alike. Not surprisingly, PBX vendors initially offered hybrid systems supporting both TDM and VoIP configuration options, whereas companies like Vonage targeted the home phone service with a pure VoIP link.

The term *VoIP* is an umbrella term that can be used to describe a whole host of competing voice technologies that were created to address the same demand. Any network protocol designed to support voice using an IP framed packet can be called VoIP. Early market offerings of VoIP sprang from vendor-specific designs long before any standards-defined protocols were ratified.

These protocols have evolved over the past 10 years as the market has grown. Initially the favored VoIP protocol was H.323 because it was the most mature evolving international "standard."

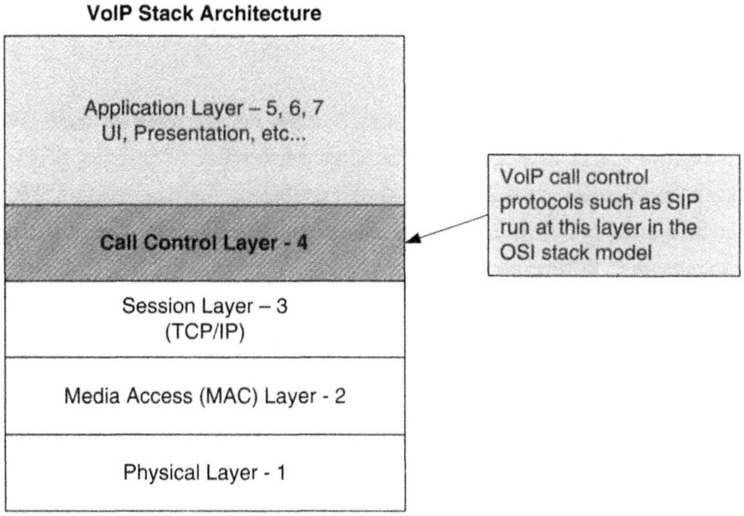

VoIP Stack Architecture

Application Layer – 5, 6, 7 UI, Presentation, etc...
Call Control Layer - 4
Session Layer – 3 (TCP/IP)
Media Access (MAC) Layer - 2
Physical Layer - 1

VoIP call control protocols such as SIP run at this layer in the OSI stack model

Figure 4.2: VoIP stack diagram.

At that time, the SIP standard was just in the initial definition stages, but today the VoIP standard of choice for UMC must be based on IETF RFC-3261: SIP.

Skype and other "soft phones" have become very popular for enabling telephony functionality on personal computers. Skype is an example of a very popular (more than 10 million subscribers) VoIP-based application that is not based on SIP. A Skype user may communicate with a SIP-VoIP user but only through their Skype-Out services that bridge such calls through the PSTN.

4.6   Hotspot and Hotzone Support

Because our mobile societies have an almost insatiable desire for access to the Internet, public WiFi services have been proliferating across every nation on the globe. Popularly known has *hotspots*, these points of wireless Internet access have been installed in hotels, airports, coffee shops, shopping malls, and municipal open areas. Commercial businesses serving the traveler see support of a WiFi/Internet service as a method for obtaining incremental revenue through an additional charge of $5 to $10 per night in hotels or through monthly subscription charges. City "fathers" have viewed support of a free WiFi service as a method for reenergizing downtown businesses in many major cities across the United States, with the most well-known municipally sponsored hotspot service in Philadelphia, where some 135 square miles of WiFi coverage has been deployed, including almost 80% of the city commercial area. Other cities are seriously considering or have deployed such services with the hope that doing so will catalyze downtown businesses.[9]

The number of global public WiFi hotspots is now in the tens of thousands and growing at a double-digit annual rate. Such an explosion of public WiFi/Internet services makes these resources an appealing candidate for use with UMC solutions. Ideally, these hotspots provide extended inexpensive wireless coverage to augment any cellular coverage. Whereas some carriers have offered a home-based FMC solution, these public services can now provide "off-campus" or away-from-home connectivity that was not possible before. Growth in this market offers some extended access for UMC users, but it does pose some challenges unique to this part of the wireless market.

---

[9] Mountain View, California; Minneapolis, Minnesota; and New Orleans, Louisiana, are just three examples of other municipal-sponsored WiFi hotspots.

Today a public hotspot is, for all practical purposes, a "crapshoot" when it comes to assurance of support for wireless voice applications. Initially designed to accommodate just HTML browser traffic, most hotpots have no QoS enabled, and whether or not the provider has considered support of real-time applications is an unknown because there is often no public posting regarding any level of QoS support. Deployment of "mesh" WiFi networks for municipal hotspots can further complicate UMC support by virtue of the fact that there are no standards that dictate QoS within a mesh "cloud." Additionally, any upstream QoS is a question, since your call audio traffic traverses the Internet, which may impact the experienced voice quality.

As the UMC market expands, services provided by hotspot providers will rise to the level demanded by the mobile market. Most of the appropriate WiFi QoS and security standards have been ratified and can be deployed at the hotspot. In conjunction, the backhaul link to the Internet will eventually be provided with the appropriate QoS so that hotspots can reliably support UMC voice quality requirements end to end. Until such time, assurance of the best voice quality over a wireless hotspot will be in doubt, and the best advice is to personally test your favorite hotspot for its specific level of QoS service.

4.7   PBX/iPBX Integration Considerations

Basic telephony functions such as call waiting, call hold, and call conferencing are givens to today's mobile phone user and enterprise phone user alike. Support for these features in a UMC solution, however, has some unique challenges to overcome. Being able to invoke a three-way conference over a carrier-centric UMC connection is not a major challenge, since those base functions are serviced within the carrier network, and with the exception of some signaling challenges (discussed in a later chapter), the UMC phone operates as usual. The challenge of providing these ubiquitous features lies with the enterprise-centric UMC products.

4.8   Network Security Considerations

Consumer UMC subscribers won't have to worry about application security the way an enterprise user will; typically they leave such concerns to the cellular provider. Enterprise or SMB users, on the other hand, have concerns about guarding their corporate data and preventing unlawful access to their corporate networks. This concern will require imposing additional security measures beyond that required for on-campus access.

Wireless access to a corporate network via WiFi often has additional security functions enforced, even when implemented inside a corporate network framework. Besides enforcing the highest-level WiFi security policy, the WiFi controller elements are often placed inside a *demilitarized zone* (DMZ). A DMZ is a segregated LAN element within a corporate network that has special security and authentication applied because of a higher intrusion risk.

Access to any IP-based network through the Internet (see Figure 4.3) is usually supported by implementation of a firewall, a Network Address Translation (NAT), and/ or a Session Border Controller (SBC) interface. These network elements all supply some level of security for remote network access to control or block unauthorized access to a network. At a minimum, some firewall service will be implemented that can block specific network addresses or protocols from interacting with applications within the corporate network. A NAT provides a convenient way of managing public IP addresses that permits hiding the internal IP address specifics while providing an access management point. A SBC is a voice-specific network device that is often configured at the edge of the corporate network to apply voice- or video-specific access control. The specific functions

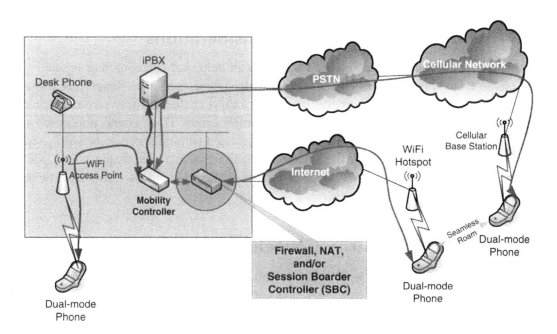

Figure 4.3: UMC hosting network security options.

of these components may be converged in some commercially available solutions that will simplify implementation of such network edge components. Specifics for each of these network edge security elements are found in later chapters of this book.

Complementing any remote LAN access security network elements, enforcement of a corporate VPN that provides a corporate-managed end-to-end secure link may also be in place. However, such security measures often limit either the configuration options of the mobile device or the locations at which these devices may access the corporate network.

Selection of any UMC vendor should also include understanding what security elements they may provide at an application-to-mobility-controller level. What about rogue device detection? Authentication schemes should authenticate both user *and* device. Do they encrypt signaling and media streams? These are very important considerations to be made by any enterprise considering a UMC deployment.

4.9   Solution Management Considerations

With any communication system implementation, someone takes responsibility for the provisioning: activation, configuration, and customization. When purchasing a cell phone from a carrier, the customer receives a unit that has been activated (the phone number registered with the wireless provider) and ready for use. The customer completes the provisioning by setting up her carrier voicemail system (with a password and so forth). It is the carrier that manages the phone from the standpoint of enabling it, activating packet data services, managing internetwork roaming, and arbitraging the billing.

Any UMC implementation will have similar deployment and management challenges. In general, consumer UMC offerings have service provider-centric management designs, and enterprise UMC offerings have enterprise-centric management designs. Because of the multinetwork nature of a UMC solution, provisioning and management tasks will have a broader functional scope and are, therefore, more complex. For this reason, understanding the underlying management design is an important factor in choosing the most appropriate UMC solution.

4.9.1   Configuration Management Considerations

The mere fact that a UMC device is mobile dictates that any updates to the firmware or configuration information must be supported wirelessly. It is important that any UMC offering support Over-the-Air (OTA) services that can update the mobile device regardless of the geographic location of that user or device. Updates by either a WiFi/ Internet or cellular packet data connection should be supported to provide a seamless experience for the user. In the case of a consumer UMC product, the hosting carrier will manage such updates, and with an enterprise product, the enterprise IT department will typically manage such updates.

4.9.2   Network Access Management Considerations

Typically the servicing wireless carrier will control an individual user's access to the cellular network (through a SIM or other means), but management of the WiFi zone access is another matter. Carrier-sponsored UMC solutions (such as T-Mobile's @ Home) will be managed by the carrier because the supported WiFi environments are well known and directly managed by the service provider.

Enterprise UMC products pose a different problem. WiFi networks are identified by their Extended Service Set Identifier (ESSID); assignment of this ID is under the direct management of the hosting enterprise. This means that for any UMC product that is adopted into an enterprise, some level of WiFi network management will be required at the end-user administration level. Support of hotspots adds another dimension of what should be managed to provide the broadest wireless access. A site UMC administrator may set up a defined set of ESSIDs that are supported by the corporation, but it is possible for an end user to add to this list through the WiFi utilities provided with each dual-mode device. For proper enterprise UMC support, it is also required that the person responsible for provisioning have access to the key WiFi security configuration information. Going beyond management of WiFi connectivity, some UMC products require site-specific location procedures.

4.9.3   Directory Access Management Considerations

To make any UMC product truly useful, there must be some support for a phone number dial list (see Figure 4.4). It is virtually impossible for any individual to memorize all the phone numbers that they might want to dial, especially on a phone that may be used for business purposes. The most straightforward method to provide a dialing directory is to integrate with the email subsystem or native dialer that may also be supported on the mobile device (for example, Microsoft Outlook CONTACTS). These lists typically reflect a user-selected list of contacts that include friends, family,

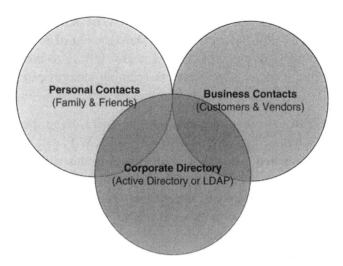

Figure 4.4: Dial list content classes.

vendors, customers, and business associates, where some of the content is privately managed. UMC devices used in an enterprise context will have a larger scope for the source of any dial list: the corporate directory.

Maximizing the effective use of a dial directory for a mobile road warrior will make some provision for merging contents from all these sources. Because corporate directories can be very large, some type of content filtering must be applied to provide access to such lists on a mobile handset, where consideration of storage capacity and performance imposes such a requirement.

Additional number sources may be found stored on the SIM for GSM phones and may be merged in the displayed dial list. Adding to the persistent dial list should also be possible by user operations of selection from call-received logs. Such numbers can be sources for new dial directory entries but must be accompanied by the ability to edit such numbers, since a fully qualified number might not be presented in this list.

4.9.4 Cost Management Considerations

For a consumer UMC solution, the main responsibility for cost control is in the hands of the user and revolves around management of minutes. In a typical plan, there are fixed costs associated with use of a phone until the allowed monthly limit has been exceeded. At this point, depending on the details of the SLA, charges will be applied on a 20–35¢-per-minute rate. Management of additional potential cellular network-to-network roaming charges is also left up to the user. For the international traveler, cross-wireless network roaming charges can be quite significant.

For the enterprise UMC user, consideration for *least-cost routing* should be handled as part of the system solution. Typically, where accessible, WiFi call "minutes" will be free or at low cost compared to a cellular call. However, this will not always be the case, and in cases where a corporation might want to emphasize cost over functionality, making a direct cell-to-cell call may be the cheapest, particularly considering a possible international call. The cost control policies of a UMC product must be considered in making an enterprise purchase decision.

4.9.5   IMS Considerations

Though not a real "buy" consideration in early 2008, the way IMS is integrated in a UMC under investigation can be a pivotal consideration. IMS is a concept of a virtually ubiquitous IP "cloud" (much like the PSTN) that is not geographically specific and provides unlimited IP-based application services and transport. Inherent in the IMS concept is the fact that access to existing wireless and PSTN resources may be accessed through the IMS interface. This means that all gateway services to these legacy types of communications will be supplied by the IMS vendor and will not be the responsibility of the individual application supplier.

The problem with IMS is that it will come to market as a vendor-by-vendor offering. Major communication companies are developing their own IMS services with no guarantee of interoperability for peripherals that interface with that IMS. Much like the interoperability problem faced by the WiFi vendor community that motivated the formation of the WFA, a similar dynamic will most likely occur in the IMS sector. Some intervendor consortium will be formed to validate and guarantee interoperability at the IMS level.

4.9.6   Support Considerations

Because UMC products will be brought to market through collaboration in a multivendor effort, support of the individual components may be somewhat fragmented.

A problem with making a UMC phone call may be a WiFi configuration issue, and the handset or mobility appliance vendor can be of little assistance in this site-specific situation. Typically, a multitier support matrix will exist, where Tier #1 (who you call first) will be the systems integrator (SI) or value-added reseller (VAR) that installed the system. Tier #2 and #3 levels fan out to include the individual component manufacturers or service providers. The handset manufacturer will be of no assistance with a PBX problem, or the mobility appliance vendor cannot help with the cellular network service problem. The important thing to understand in making a UMC purchase decision is to be sure the "support" service features are clearly outlined as part of the agreement. A concise troubleshooting flowchart should be provided to aid in portioning the point of problem. In most cases, service for a workday (9:00 a.m. to 5:00 p.m.) or a full day (24 hours) will be offered as a recurring charge and bundled with the original sale.

CHAPTER 5

UMC: Current Market Solution Overview

This chapter describes the major solution architectures (enterprise- and carrier-centric) and discusses the design approaches of each and the state of commercial offerings. There are distinct differences in the requirements of each market segment but a common functional intersection of seamless roaming across the disparate wireless networks.

5.1   Market Drivers for Mobile Communications

One factor driving the demand for a more ubiquitous mobile communication system is the people of *Generation Y:* the 20- to 30-year-olds who prefer use of a cellular phone over use of an old analog "home phone" as their primary telephony service. In addition, statistics indicate that the total available market (TAM) for cellular phones in most industrialized nations is now approaching saturation, and this growing market sees value in extended mobility, which sparks the UMC demand. Coverage deficiencies in standard cellular network solutions now drive the demand for a UMC solution—a virtually ubiquitous wireless service accessible anywhere, anytime.

Another factor that contributes to UMC demand is the fact that the proliferation of WiFi has permeated both private and business lives of most people. The cost of a wireless router and a monthly Internet service is within reach of all but the impoverished. Businesses now also see the benefit of deploying WLAN technology as an extension to their corporate LANs to give them the in-building mobile flexibility they need to do business.

The final factor impacting UMC demand is the availability of dual-mode handsets that now better meet the UMC functional requirements. Cost and battery life could still be issues, but devices are now commercially available that provide excellent voice quality in both WiFi and cellular environments.

All the "planets are aligned" as the necessary solution components to support UMC are commercially available. What is the state of readiness of these solution components? Are there still feature deficiencies or mere deployment inadequacies?

5.2   Cellular Solutions Cover All Outdoors

As discussed in previous sections, the major flaw of most cellular networks is the lack of complete coverage that includes homes and in-building service for offices. Further complicating a cellular provider's goal of providing a total mobile solution is the fact that packet data service is often spotty or nonexistent. For example, in the U.S. "breadbasket" region (Iowa, Kansas, Montana, Colorado, and so on), CDMA is the predominant cellular service but may only be supporting the 1XRTT packet service. Because this was a first-generation service, it has neither the bandwidth nor the data rate to reliably support any kind of mobile phone function. Some UMC solutions will invoke a cellular-based call via packet services, and in the case of 1XRTT, these attempts regularly fail because of the latency inherent in the system; so the preferred network should support something beyond 1XRTT. Even with second- and third-generation packet services such as GPRS or EDGE, this coverage may be spotty even in some urban areas, and any UMC solution that depends on carrier packet services could have an Achilles' heel in terms of not being able to provide consistent and reliable service in certain geographic locations.

Expecting a UMC application to work the same way in all possible geographies is not a reality, even in 2008. It is important to know the nature of the cellular coverage (both voice and packet data) from your provider of choice.

5.3   WiFi Solutions Cover Indoors

Even in 2008, many deployed commercial WiFi networks fall short in terms of being able to adequately support good voice quality for UMC-type applications. Where WiFi coverage is under the control of a business, good overlapping coverage can be mandated as part of the WLAN service contract. This may be more expensive, but it's critical for realizing consistently good WiFi voice quality within any one WLAN. Additionally, the level of security imposed may have a negative impact on voice quality, and the deploying business may have to compromise between high security and good voice quality over the WiFi segments.

Public WiFi resources are typically not deployed with voice support in mind, and the individual user has little control over the resulting experience. Unless the hotspot or municipal WiFi provider has specifically designed the network for support of real-time (voice or video) wireless applications, the voice experience may be random and intermittent. Tens of thousands of public WiFi hotspots are available, but for the frequent hotspot user, it would be a good idea to characterize the voice experience at the most frequented places. Even then, the imposed QoS might not be sufficient to maintain good voice quality in the face of large file downloads.

Any shortfall in support of availability or realization of good voice quality will be made up over time as the market matures. Just as the first televisions introduced were black and white with small screens, the market evolved over time to the HDTV commercial offerings we have today. However, without the early sales of the black-and-white TVs and ongoing success in the television industry, there would never have been a possibility of HDTV.

5.4   UMC Carrier-Centric Architectures

Consumer-targeted UMC solutions fall into a class of products with a carrier-centric design. Since it is the wireless carrier that provides the mobile features, it makes sense that the control point for these features would reside inside the carrier network. By extending the perceived network coverage to include WiFi, the carrier provides a more robust and highly available network service and will help minimize subscriber "churn" for the carriers. Success of any carrier-centric UMC solution not only requires the end user to purchase a dual-mode handset, but correspondingly, the carrier must upgrade its networks to accommodate this new feature, which can impose a regional deployment dependency when rolling out such a product.

Two different technological UMC solutions have evolved from the 3GPP standards efforts that are targeting the same end user. Each has a carrier mobility component required to be installed in the wireless infrastructure, but the solution approach for the way the client accesses these services is quite different.

5.4.1   Universal Mobile Access

The underlying architecture of the Universal Mobile Access (UMA) design is one in which the GSM signaling and audio streams are tunneled over an IP transport in WiFi coverage. This is accomplished by the client establishing an IP link with a corresponding UMA point-of-presence network element called a UMA Network Controller (UNC). To the cellular network, such devices appear to be functionally identical to a standard cellular phone. Because of the collaboration between the UNC and the handset client, the functions supported at the handset are identical, regardless of whether they are in cellular or WiFi coverage (see Figure 5.1).

The UMA architecture requires a dedicated handset that will forward all traffic (voice or data) back through a WiFi connection into the hosting cellular cloud to the other phone or service point. The call is "anchored" in the carrier cloud, and when a user exits a WiFi coverage area, the UNC passes the call to the neighboring cellular base station, just the way a call is passed between base stations when you're driving in a car. No user action is required to enable the transition. Roaming into a new WiFi coverage area will cause the network to identify a nearby UNC where a WiFi transport call may be established, and the bearer channel call will be terminated.

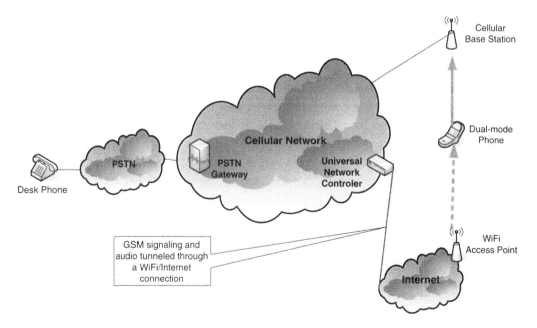

Figure 5.1: UMA cellular-centric FMC solution.

Another implementation of the UMA architecture involves support of single-mode cellular phones with femtocell technology. This UMA/femtocell hybrid leverages a cellular provider's investment in 3GPP-GAN to maintain better end-to-end control over the calling environment (see Figure 5.2).

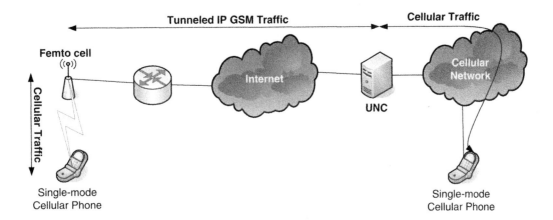

Figure 5.2: UMA/femtocell hybrid.

Regardless of the specific wireless implementation option deployed, from the user's perspective the mobile phone continues to support the same set of carrier-provided features, regardless of whether the user is in WLAN or carrier network coverage. Once the dual-mode phone is properly configured with an acceptable WiFi network profile list, the UMA user has freedom to roam in and out of WiFi coverage with the guarantee that the phone call will not be dropped. The behavior of a single-mode phone can, unknowingly, also take advantage of the UMA technology where femtocell hybrids may be deployed.

For the Windows desktop user, Vitendo recently announced its vFone 4.0, the first UMA-based softphone for PCs. The vFone softphone mimics the behavior of a GSM phone and allows the user to make and receive calls and send SMS messages from laptops or desktops. This device is a first in the market where a *fixed* desktop application emulates a *mobile* application—kind of *reverse*-convergence. Use of the cellular network services will require some new class of SLA that is to be defined between the Vitendo team and carriers.

5.4.2   Voice Call Continuity

A voice call continuity (VCC)-UMC solution is much like a UMA solution in that the call is anchored in the cellular network, but the major difference is that the VCC client (see Figure 5.3) is based on the SIP standard and not a GSM cellular phone emulation model. This approach has advantages in its compatibility with iPBX solutions and IMS networks, since most of these products "speak" SIP natively.

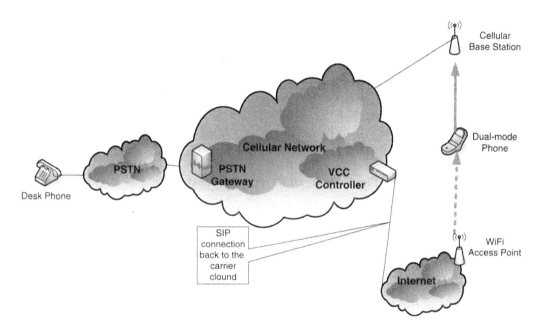

Figure 5.3: VCC cellular-centric UMC solution.

This approach, in a carrier context, does have drawbacks not found with UMA where many carrier-derived services are not supported. Features such as messaging and Push-to-Talk are not possible with existing VCC offerings. This means that only baseline carrier services are supported with VCC solutions and the end user must make a purchasing value decision regarding the presence or absence of certain key features. VCC's industry focus is now linked with IMS deployments, which will pace any market adoption of this design pending rollout of IMS networks.

5.4.3   General Carrier UMC Model

There are feature differences between UMA and VCC approaches that will drive the market to eventually accept one or the other. As implemented to date, the user experience in terms of phone calls is pretty much paired.

Figure 5.4 is an example of the way the connections are managed in a 3GPP carrier-centric solution in a phone call between a mobile handset and a landline phone. The call is initiated by the mobile handset when in range of a hotspot WiFi connection (1) where the audio and signaling are routed through the WiFi access point and through the Internet to a carrier mobility controller that emulates a standard cellular base station. Each variant from different carriers may have some small differences, but the basic model is simple emulation of a base station, which makes the WiFi handset appear as a standard cellular phone to the carrier network. The mobile user will eventually move out of range from the hosting WiFi service (2), and the mobile phone will roam to a standard base station using the GSM or CDMA radio functionality. Neither the mobile user nor the fixed-line subscriber is aware that the call was rerouted across two different wireless networks.

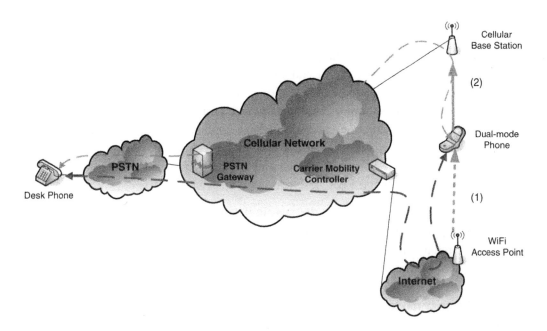

Figure 5.4: Carrier-centric roam sequence.

UMA support has already been announced and deployed by major carriers and handset manufacturers, with little or no announced support for VCC. Given the fact that UMA provides the UMC network roaming agility and unrestricted access to the carrier phone services, UMA appears to be the winning UMC solution for consumers going forward.

5.5 UMC Enterprise-Centric Architectures

A UMC enterprise-centric solution is designed to provide enterprise-critical mobility features and be hosted by the enterprise. At its core, the enterprise UMC solution presumes that the WiFi and wired network services are a single logical resource under the control of the enterprise, with application control being retained within the enterprise IT context. Such a design has no innate carrier dependencies and allows for a more "native" architecture to be implemented where Voice-over-IP (VoIP) can be supported as part of a solution complementing the mobility capabilities of the WAN component and iPBX VoIP solutions that may be deployed. The carrier-centric solutions, on the other hand, take the inverse view that WiFi networks are logical extensions of the carrier wireless network and merely act as a transport for the cellular traffic, with application control remaining with the carrier/service provider network. Because of the diverse perspectives on how such resources are provisioned and managed, functionally distinct solutions will emerge.

The rapidly growing enterprise adoption of WiFi (802.11 WLANs) and VoIP as well as the availability of dual-mode (WiFi and cellular) devices are the key business and technological events that have catalyzed the enterprise UMC demand. The benefit of UMC is that it leverages these technologies (and investments) to support important mobility functionality while the enterprise retains application control. Additionally, integration with the enterprise PBX/iPBX is a unique value-add, resulting in a single-number reach and single voicemail design and telephony functional integration of the mobile units with traditional PBX desk sets.

Figure 5.5 depicts an enterprise call that was initiated within the corporate campus WiFi (1) to a desk phone, which then roams out into a nearby public WiFi hotspot (2). The call is anchored and managed through the enterprise premises mobility controller. From the public hotspot, the user roams out of range and is automatically roamed into the cellular service (3). Through all these transitions, the call continuity is sustained and, unlike the carrier model, enterprise monitoring and use/security policies may be applied. Though the sequences may look confusing and somewhat complex, the end user experiences a seamless voice experience, regardless of network proximity.

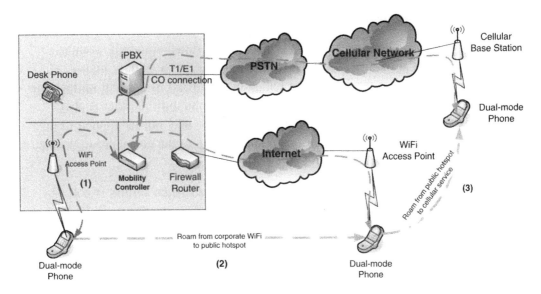

Figure 5.5: Enterprise UMC solution.

The distinguishing proposition of an enterprise-centric UMC solution is the fact that the call control point is resident within the corporate network. UMC handsets will be viewed by the enterprise as logical nodes of their corporate network and extensions of their PBX with a definable business value. To achieve broad adoption in the worldwide corporate market, any enterprise-centric UMC solution will have an underlying *agnostic* design perspective and will *ideally* be agnostic with regard to available cellular services, supported WLAN product, and hosting handset environment, because the Fortune 2000 enterprises are so fragmented due to their evolutionary adoption of each of these technologies.

5.6   ROI Models and Solution Trends

Besides the thrill of not having your call dropped when you enter a building, what are the criteria for purchasing a UMC solution? From the rush to purchase Apple iPhone sales, you might think that simply the technology intrigue and the "wow" factor are sufficient to make a buy decision. Such customers most certainly can be classed as *early adopters*. These are the foolhardy ones who pursue the latest of everything, for the intellectual and trendsetter thrill. But what about after the initial surge of early adopters? What makes the next wave of buyers purchase a UMC solution?

5.6.1   UMC Consumer Solutions

For the current generation of UMC cellular-centric consumer solutions, seamless connectivity comes at the price of the dual-mode phone and a slightly higher monthly charge. Whether the cost of the phone is subsidized or not will mitigate the actual price of the phone ($400–800 list). Typically, a monthly flat rate is also applied to calls from a WiFi zone, a charge that may or may not be applied to the monthly allocation of minutes. But at this point, it is up to the consumer to decide whether or not this is a value worth the monthly charges.

For UMC consumers, the hope of free (or cheap) long distance calling is not realistic. Tariffs on long distance calls made from a carrier-sanctioned WiFi zone have no impact on lowering any talk-time charges. The lower rates offered by some residential VoIP providers are not part of the package offered to the UMC user.

For many individual consumers, selecting a UMC solution is not driven by cost but is rather a matter of personal convenience or personal situation. For example, a multistory apartment complex might not provide adequate cellular coverage for the interior apartments. Having a low-cost wireless/Internet connection can allow a Generation Y person to have that seamless wireless experience with the dual-mode cellular phone she desires and avoid installing a fixed-line phone. Most certainly, the extended "accessibility" goes further toward encouraging on-the-go social networking.

5.6.2   UMC Enterprise Solutions

The whole UMC buy decision process for an enterprise or SMB is completely different. Because such decisions impact the success of businesses, they are evaluated in much the same fashion in considering any technology purchase. Return on investment, or ROI, becomes a real consideration in this process and will include estimates of capital expenditures and long-term operating costs. Depending on the ROI considerations a business makes, ROI factors can be classed into two categories:

- *Hard ROI.* Actual capital expenditures and detailed cost of operations.

- *Soft ROI.* Performance and productivity enhancements derived from the UMC solution.

Because of the different implementation architectures, enterprise-centric UMC products can have radically different ROI characteristics and considerations than the corresponding consumer products.

With the enterprise UMC solutions, the promise of free (or virtually free) phone calls can be met. This is because the system call control point is within the enterprise network and is totally decoupled from the paired cellular service. Such architectures allow *pure* SIP-VoIP calls to be routed between two iPBX managed phones without having to traverse the peer cellular network. As long as the phone stays under WiFi coverage, the call remains on a pure, enterprise-managed IP network and costs virtually nothing. Only when the UMC phone exits WiFi coverage will the services of the cellular network be invoked to support the call.

The fact that not all calls will have to traverse the wireless network means that users may have a reduced monthly minute requirement for their cellular services. Since some analysts have reported that some 60% of business cell phone calls were made inside a business that was covered by corporate WiFi, this means that an optimistic 60% reduction could be projected for some individual UMC users. Though this example would not be common, there will be a reduction in cellular minutes used any time the WiFi services are accessed, which will reduce the average cellular mobile minutes required by any one user. Such a potential reduction in required cellular minutes can allow the business to negotiate a lower number of per-person minutes in the SLA or could mobilize more associates with the same budget.

A secondary benefit available to enterprise customers is the fact that many of the dual-mode, UMC-compatible handsets can be platforms for other job-related mobility application functions and can afford a convergence (or reduction) in the number of devices per associate. Many road warriors carry a mobile phone, a pager, a cellular phone, and some other computing device. With the current generation of UMC-class devices, support for mobile email, telephony, Push-to-Talk, instant messaging, presence, and vertical market applications is possible with a single multifunction device.

One additional benefit of an enterprise or SMB UMC solution is the productivity gain. Though it is sometimes difficult to equate to hard dollar savings, it is fairly easy to understand such a benefit. The first productivity gain is by virtue of the mobile user being more accessible, which facilitates quicker responses with key personnel. Eliminating or minimizing having to answer voicemails has perhaps the greatest positive impact on productivity. When a mobile employee has a deskbound phone, returning to his desk to pick up and return voicemail can cost him 5–10 minutes per call. A certain percentage of this time is, therefore, not lost if he can take the call in real time, when he is away from the desk. Telephone "tag" further complicates this matter and further erodes productivity. A good example of the impact of accessibility is if a mobile worker who earns $80,000 a year can avoid just five missed calls per day, the equated conservative annual time savings will be more than one week of that worker's time. If calculated on the annual cost, the value loss of this worker is more than $2,000. Most UMC systems would pay for themselves in less than six months if the "soft" ROI were considered.

5.7   WLAN/Internet vs. Cellular: A Commercial Battleground

It seems clear that the trend for adoption of VoIP continues to gain momentum and that the death of the faithful PSTN is predictable. Such a strong statement may be somewhat shocking, but as the Internet permeates the personal and professional lives of industrialized nations, the need for the old circuit-switched communication technologies begins to wane. Not only have all major international PBX vendors made the commitment to VoIP, but the scope and availability of the Internet in homes and offices provide a natural replacement technology for the old analog home phone. But is there turbulence in the marketplace as the major players vie for position?

Three major technology and business forces struggle for survival and prominence in this new wireless mobile world:

- *Wireline providers.* Traditional circuit switched (fixed) network vendors.

- *Wireless network providers.* Existing cellular network providers.

- *Nuevo communication network providers.* New competitors with mobile solutions and those supporting national and international access to IP-based IMS services.

5.7.1   Wireline Providers

This major business segment has the most to lose with adoption of wireless and VoIP technologies as the preferred communication modes. The legacy circuit-switched network that has provided the world with reliable telephone service for over 100 years is now threatened by the disruptive packet-based technology. The wireline subscriber customer base declines every year as individuals and companies disconnect from the PSTN and deploy one of the new alternatives. Even without the threat of VoIP, large segments of the wireline subscriber base have moved over to a pure cellular-based service.

What wireline vendors are doing in the face of this threat is to integrate the new technologies to offer a "hybrid" service or merge with vendors already in the competitive technology space. The PSTN will be with us for a long time because it is so pervasive and still provides reliable telephony services to large subscriber bases. However, the handwriting is on the wall: Wireless and VoIP services will be the dominant communication services in the 21st century.

Many of the Baby Bells merged with businesses that also provide cellular services, which makes a lot of business sense. Being able to offer both wireline and cellular services means that revenues are not lost but rather balanced between the two business divisions, thus stabilizing the overall business base of the whole company. AT&T Wireless merged with SBC in 2005[1] to form the new AT&T, which provides both classes of service. The irony is that we are seeing a reverse of the 1984 breakup of the Bell System as these same entities converge due to this radical change in the business environment.

Another threat to the wireline providers comes from the cable industry. As an additional connection to the home (and office), support of telephony services now complements the traditional television and Internet services provided by these businesses. Comcast, the nation's largest cable provider, now offers the triple play—television, Internet, and telephony—as part of its services. It is a small step for the cable industry to negotiate a mobile virtual network operator (MVNO) contract with a major carrier. An MVNO allows a third party to broker services from a national wireless carrier and extend its basic product offerings to include wireless network subscriptions. Wireline providers may also become MVNOs, but it will be no surprise to see cable companies eventually offer a full range of communication services in direct competition with wireline providers.

---

[1] AT&T also merged with Cingular under the AT&T banner.

5.7.2 Wireless Network Providers

The highly successful wireless carriers have a different play in meeting the UMC opportunity. As an industry, they have stepped out and defined two separate solutions to provide seamless mobility:

- *Femtocells or picocells.* These are IP-connected pseudo-base stations that are designed for indoor installation. With these deployed, the cellular network coverage is extended inside hosting facilities (corporations, malls, airports, hospitals, and so on).

- *3GPP-v6 UMA standard.* The latest ITU communications standards coming from the 3rd Generation Partnership Project (3GPP) include carrier-centric solutions to support seamless roaming across WiFi and cellular networks.

Femtocells and picocells fulfill the same function of emulating a cellular base station but are designed to be installed inside a building. The prime difference between these two product families is the cost, capacity, and comparative RF range. Typically femtocells are designed for the home or small office/home office (SOHO) market and cost between $100 and $200, with a range of about 300 feet. Picocells, on the other hand, cost around $2,000 and have a 100–200-meter range with a higher bandwidth capacity.

The challenge with the femtocell/picocell strategy will be not one of technology but one of channel engagement. Because of the relatively low cost of a femtocell, sales to the home market will be more straightforward and possible through existing retail channels. The question facing the success for picocells is, who will pay for the enterprise installation capital expenditures? These more costly carrier solutions are targeted for markets with higher bandwidth demand. There would be little or no ROI rationale for a single enterprise installing such a solution, because it could rely on the wireless WiFi infrastructure that was installed as part of its corporate network. Most likely what will happen is that the carriers themselves will fund deployments of picocells in very public sites such as shopping malls, airports, and public buildings. Several national carriers have announced their support of femtocells,[2] but the jury is out on how successful this product will be in either the home or the office.

---

[2] Verizon and Sprint/Nextel have announced their support of femtocellular technologies.

What about support of VoIP over CDC? With the second- and third-generation packet services offered by the cellular providers, there is no technical obstacle for implementing this approach. Using the IP-based transport services from the carrier, any SIP-based softphone could be supported with virtually no validation testing. The problem with this approach is a congestion and competition one faced by the cellular providers. Most cellular providers view VoIP as a competing technology and a threat to their business. VoIP over CDC faces latency problems in assuring good voice quality, but more important, it also poses congestion problems for the carriers that would impact service provided to subscribers for World Wide Web access. For this reason, many carriers have implemented packet inspection logic at points of presence that blocks attempts to use VoIP over CDC. Additionally, they have put pressure on many dual-mode handset vendors to eliminate or minimize the ability to run UMC applications on these devices. Typical of such policies is withholding internal features that will cripple UMC applications.

5.7.3 Nuevo Communication Providers

New players or old players with new product offerings are coming to the market to capture part of the UMC business. Companies that have never been in the telecommunications industry are now announcing their entry into this arena. The advent of WiFi and VoIP technologies opens the door for new competitors to enter the telecommunications market. Cisco Systems, which owns the lion's share of the enterprise networking business, also offers WiFi and VoIP solutions as part of its offerings. There is strong indication that the company will move into the UMC arena, based on several recent company buyouts. Another natural UMC provider is the PBX vendors themselves. With the focus on VoIP as the technology of preference, these companies must offset any potential desk-set revenue losses by offering a corresponding wireless product. Most of the key PBX vendors have already announced support for cellular phone integration (e.g., Avaya's Extension to Cellular) or partnerships with Nokia for its dual-mode Eseries phones.

Perhaps the player to make the biggest impact on adoption of VoIP and UMC products will be the IMS vendors. An IP Multimedia Subsystem (IMS) is an architecture that leverages the Internet's pervasiveness and adds a service layer focused on supporting IP-based applications of all kinds. On such an infrastructure, applications of almost limitless functionality can be hosted, including telecommunication applications that span diverse network types.

IMSs will emerge on the market in 2008, but will be hosted vendor specifically. In a market where multiple competitive IMS services are available, there may be some ambivalence on the part of application developers to select one on which to launch their products. It is assumed that some incompatibilities will be encountered anytime an application needs to traverse multiple IMS networks. All this complexity and the way it impacts the UMC business are to be determined.

CHAPTER 6

UMC: Layer 1 and 2 Media Requirements

The key component of any telephony architecture is the media—that is, the methodology used to digitize a human voice and transport it over a network, to be converted back into an analog signal to be listened to.

6.1   A Bit of History

Original telephony products were based on an analog design in which a human voice was changed into a varying voltage or current that was translated at the other end in an inverse manner. Typically requiring relatively high voltages (48 volts DC), these devices "worked" for over a century but at a price of heavy (literally) wiring infrastructure and high power consumption. In a circuit-switched environment, however, this was pretty straightforward, since the end-to-end connection became a physical wired connection between two analog devices. To select the person you wanted to talk to, you had to provide an operator with a number (verbally); the operator would then patch two wires together to complete the physical circuit. Automated circuit switching was invented to simplify this operation through use of pulse dialing and, later, dual-tone multiple frequency (DTMF) signaling for destination identification. Digital interfaces were provided as the telephony network evolved, but they remained, basically, a circuit-switched network.

Late in the 20th century, three disruptive technologies came on the scene that revolutionized the telephone industry:

- Cellular telephony

- Voice-over-IP telephony

- Wireless LANs

6.1.1   Cellular Telephony

Following the path of its circuit-switched predecessor, cellular telephony began life as an analog service: Advanced Mobile Phone Systems (AMPS). It was the wireless peer to the services provided by the PSTN, and a whole new industry was born around this technology that became pervasive in most industrialized countries. As popularity of wireless telephony grew, a trend began to emerge: The young and mobile in the societies tended to prefer to use wireless telephone services exclusively, sparking the ongoing rivalry between wired (fixed) and wireless (mobile) telephony vendors. Gateways were installed so that a fixed phone could call a mobile phone, and vice versa, but the fact remained: A fixed phone was fixed and a mobile phone was mobile. The two had different features and were supported by completely different industry vendors.

Analog wireless was eventually to be replaced by digital wireless technologies (Digital AMPS, GSM, and CDMA), resulting in a more reliable service with better voice quality experience for the mobile user. For the most part, digital wireless services have all but replaced analog wireless, but there remained a great technological gulf between the fixed telephone and the mobile phone.

Regarding the media on both analog and digital wireless telephony, it would travel from phone to base station to central network and finally to destination phone as a temporal circuit managed by the network. Roaming decisions between base stations were the responsibility of the "network," which assisted in managing congestion problems. Audio-encoding schemes and error-handling techniques were developed to improve voice quality and reliability, but the underlying media architecture remained basically the same.

6.1.2   VoIP Telephony

Digital wireless telephony was still circuit switched, and the concept of commingling voice, video, and data over a packet network came about with the maturing of the

Internet. Packetized voice was a radical idea that was met with skepticism when first proposed. As was true of so many earlier disruptive technologies,[1] many said it "couldn't be done."

However, if we look back over the past seven to eight years, we see that most of these concerns were ill founded. VoIP, as a standalone technology, has been broadly accepted and is a mainstay in the telephony industry. The market message is clear: Packet-based voice is the future; circuit-switched voice is the past.

The VoIP media, in the UMC context, traverses both wireless and wired networks and may have to conform to multiple standards for different Layer 1 and 2 technologies as it travels from caller to callee. Encoding (codecs), encryption, packet prioritization, and packet loss schemas may all be applied at different points in a single packet's travel from start to finish just playing a few milliseconds worth of audio. All of this may sound complex, but today's product offerings are up to the challenge.

6.1.3 Wireless LANs

While the original 802.11 wireless LAN design did not encompass the idea of supporting real-time applications like voice, early on, there were proponents of using this media as an extension to the hardwired media for VoIP. Once the enterprise hurdle of basic security was addressed, multiple vendors entered the market offering WiFi voice products that used proprietary solutions to obviate the QoS and bandwidth problems of the original standards-based products. Market demand for mobility then drove the standards efforts to evolve the 802.11 standard to acknowledge the needs of real-time applications such as voice and video. Today, virtually any application running successfully over Ethernet can be run over WLAN.

6.2 Cellular Networks

The evolution of cellular networks is usually spoken of in terms of *generations.* Generation #1 was a pure analog-based solution, and each subsequent generation was digital, with additional value-add functionality. Because upgrading an existing network is very expensive, moving from generation to generation was scheduled by regions and driven by potential revenue gain. High population areas were typically the first to reap

---

[1] There were skeptics with regard to 10 Mbps Ethernet over unshielded twisted pair and 100 Mbps Ethernet, to name a few examples.

the benefits of the new generation of features. Supporting these new features in rural areas was often years behind the initial urban rollouts. Also, competitive wireless network offerings might have different dominance by region and would affect carrier selection options.

The market landscape of the cellular market is diverse, depending on geographic location and technology. Differing cellular provider policies (not technologies) have major impacts on how UMC solutions can be integrated. Primarily centered on the concept of "locked" phones, cellular providers have used this technique for maintaining control over how their network can be used and who can access it. As practiced in the marketplace, the radio in the phone can be locked to a specific wireless carrier and a phone can be locked to a specific wireless carrier (GSM or CDMA). For example, with a GSM phone, the SIM is the critical component or "key" for accessing the carrier network. Since they are issued by the carriers, these components can be programmed to only associate with that carrier's signaling. Additionally, since the carriers usually certify and sell the handsets, these units are also locked to the carriers; placing a foreign SIM in a phone will fail, even though each component would technologically match the source carrier's network.

Because of market demand and the enacting of new governmental regulations, the practice of locking phones is passing. In Europe, unlocked phones may be purchased today on the open market, whereas the North American market has not completely opened up. The consumer benefit realized by unlocking SIMs and phones is that a more competitive market is energized where any SIM and any phone can now be used on any technologically compatible wireless network. No longer will a customer be forced to purchase a new cellular phone if she changes wireless carriers.

Another cellular UMC consideration is the tariff structure that is found in some regions. Because of some country-specific regulations, there can be a heavy tariff imposed on a call that traverse the fixed (PSTN) network to the wireless networks. Any path for a phone call that would route it across multiple fixed and wireless networks would add to the cost burden. An intelligent UMC that is supportive of least-cost routing will adjust the call path accordingly. To facilitate such implementations, a market of IP-to-cellular gateways has evolved to bypass the PSTN when integrating with a VoIP solution.

The following sections provide a brief overview of the features provided by the major *generations* of cellular service. It is important in selecting a UMC solution that the end customer understand the functionality and flexibility provided by each network generation, because certain UMC benefits hinge on services provided by the wireless carrier's network.

6.2.1   2.0G and 2.5G Feature Support

The second-generation cellular service was characterized by deployment of multiple competing technologies that were focused on voice traffic. Both analog and digital wireless networks were deployed based on different TDMA or CDMA schemas. In North America, Nextel utilized a proprietary TDMA design from Motorola, Verizon was based on a CDMA architecture, and AT&T (Cingular) was based on the GSM standard. In Europe, GSM was the overwhelmingly dominant technology deployed for cellular service and was rapidly being adopted worldwide as the preferred cellular technology.

Evolution to a 2.5 generation occurred primarily through adding packet data services. This allowed for Internet access and more extensive IP-based application support from mobile phones. Providing value-added extensions such as messaging, ringtones, three-way calling, and others further characterized this intermediate cellular generation.

6.2.2   3g: 3GPP/3GPP2 Feature Support

The third cellular generation is being deployed worldwide and is characterized by increased bandwidth capacity and significantly higher packet data rate capabilities. Working with international regulatory bodies, more RF spectrum was opened up (for a price), which added to the overall capacity and deployment flexibility for these networks. Significantly higher data rates were possible, with up to 384 Kbps when mobile and 2 Mbps for stationary systems.

With the 3GPP new versions, the first FMC (UMC) standards were incorporated; version 6 annotated the UMA (also known as Generic Access Network, or GAN). The two hurdles that would pace acceptance of these technologies were availability of dual-mode phones at the right price and feature support and upgrading their wireless network to support these enhanced features. As of the second quarter of 2008, only regional pilots of these services have been deployed in North America or Europe on the 3G networks.

6.2.3   4G: A Network Coming to Your Town Soon?

The future holds a lot for the wireless user: voice, video, and messaging anytime and anywhere. Though the specification has not been finalized, the 4G standard promises to be the ultimate for IP-based communications. Virtually unrestricted and untethered communications to anywhere on the face of the Earth at data rates as high as 100 Mbps and 1 Gbps are predicted.

How fast we will "see" the 4G benefits will depend on the business dynamics of identifying who will buy it and how it will be paid for. When 4G is a reality, it will most likely come intermingled with IMS and support variants of UMC as peripheral *application services*.

6.2.4   Other WWAN Technologies

For historic completeness, it is important to note that other wireless technologies have been commercially available and at one time were promoted to fill specific market needs. Most of these services are now obsolete and are not considered for any UMC application use:

- *Cellular Digital Packet Data (CDPD)*. An early AMPS cellular packet service that utilized unused bandwidth. The data rates were slow and somewhat unreliable and have been replaced by the 2G and 3G packet services.

- *DataTAC*. A wireless packet data developed by Motorola and was commercially available over the ARDIS network. This dedicated wireless network was used primarily for vertical market peer-to-peer application data transfer.

- *Mobitex*. A wireless packet-switched network based on an international standard deployed in North America and Europe. Used extensively by public safety organizations (fire, police, ambulance) in the late 1980s through the 1990s.

6.2.5   Femtocell/Picocell Technology

Femtocells and picocells are a relatively new option to extend cellular network coverage. Media flow between the mobile handset and one of these "cells" is identical to that which flows between the handset and a standard cellular base station except there is an IP transport segment in between the "cell" and the cellular "cloud" (see Figure 6.1). The cellular bearer traffic is tunneled in an IP packet as it traverses the IP segment. The features supported by this approach are identical to those provided with a UMA solution, but there is no need to purchase a new cellular phone.

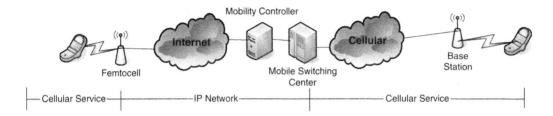

Figure 6.1: Femto/picocell topology.

Deployment of femtocells and picocells have a rosy future, as predicted by analysts, growing tenfold per year for the next two years,[2] but it may be hampered in a business context by competition from free WiFi services that provide competitive transport media.

6.2.6   UMC/Cellular Readiness

The big question regarding UMC solutions is how well they work with today's cellular networks. The answer is that most current and proposed solutions will work fine on 2.5G and 3G networks, with minor feature restrictions on some features running on the older networks. Slower data transmissions and possible carrier-specific limitations may be imposed for noncarrier-sponsored UMC solutions. Networks that are 2G will not be good networks for servicing UMC (or FMC) applications. Understanding what carrier services are available in a geographic user area is an important deployment consideration.

6.3   WLAN/802.11 Networks

The IEEE 802.11 standard has been embraced by industry and businesses as a solution to their in-building network access mobility problem. Initially, the benefit was simply to eliminate the requirement of running an Ethernet cable to a temporal location, but quickly the technology was found to be able to meet the network mobile access requirements of business associates, allowing them to roam throughout a company office being "portable," remaining connected to their corporate Ethernet network. Like

---

[2] Sanjeev Verma, cofounder of Airvana, Inc., projects the femtocell market will grow quickly, with millions installed by late 2009. Infonetics projects the annual femto/picocell market will grow to $630 million by 2010.

any technology, early products were expensive and with limited intervendor interoperability, though all WLAN vendors claimed to be compliant with the 802.11 standard. This wasn't much of a problem because there were only a few vendors in this space and you only had to ensure that you used that vendor's products.

As it has evolved, 802.11 products have several "flavors" that vary in data rates, assigned spectrum frequencies, and media encoding techniques. The 802.11a and 802.11b specifications were ratified as standards at the same time, but the first widely successful market offering was 802.11b,[3] which used the 2.4 GHz ISM spectrum and afforded raw data rates between 1 Mbps and 11 Mbps. Support for the promised higher data rates of 802.11a lagged behind 802.11b because of technical difficulties with the silicon chip circuit design. Aside from a massive problem with the initial security specification (discussed in a later chapter), WLAN vendors quickly shifted from proprietary designs to support the new standard. Competition between vendors and commercialization would energize the whole market and eventually cause prices to fall as shipment volumes rose.

As 802.11b products entered the market, it was also apparent that the 2.4 GHz band was getting crowded and had a high probability of interference from Bluetooth, Zigbee, cordless phones, baby monitors, and RFID devices to disrupt its operation. The logical move was to commercialize the standard that had less potential for interference: 802.11a. 802.11a operates in the 5.2 GHz band and provides all the same basic features of the 802.11b standard, at higher data rates. One unique feature of the a standard is that it does offer more logical channels, and that goes a long way to minimizing co-channel interference and maximizing overlapping coverage within a facility. Once chip design problems were addressed, 802.11a products began to emerge on the market in early 2001.

The popularity of 802.11b propelled the WLAN market ahead, but like most things in technology, the *status quo* speeds (data rates) could not satisfy the market, and the 802.11 working group defined 802.11g.[4] 802.11g still used the 2.4 GHz spectrum but now provided for data rates up to 54 Mbps. Commercially, almost all WLAN products now provide transparent support for dual-mode 802.11b *and* 802.11g operation.

---

[3] The standard specified 1 and 2 Mbps frequency-hopping WLAN modes that were commercially available from a few vendors.

[4] Why 802.11g? The intervening alpha characters between 802.11*b* and 802.11*g* were already assigned to other 802.11 working groups.

Near the end of 2007, the WLAN industry rushed out the latest variant: IEEE 802.11n. This standard has not been ratified, but industry players have worked with the WFA to define a certifiable "prestandard" version of this new wireless technology. The appeal of n is its spectrum flexibility and increased data rates (up to 300 Mbps), among other things. The industry has high hopes for this latest WLAN offering, but success will be defined by the market voting with dollars. Whether or not n mobile handsets will appear is a question discussed in later sections.

6.3.1 Common WLAN Voice Problems

Regardless of which flavor of 802.11 is deployed, there are several problems that need to be considered for support of a WiFi voice application like UMC:

- *Load balancing*. Too many real-time applications utilizing a single AP will saturate channel capacity and therefore deteriorate voice quality, regardless of QoS.

- *Overlapping coverage*. Inefficient RF overlapping coverage for each access point will impact voice quality and even jeopardize call retention when roaming between access points.

6.3.2 Load-Balancing Considerations

The ability to manage AP congestion from a network perspective is important in a fully loaded WLAN. If all UMC users initiate calls while in a single locale (a meeting room), there will be bandwidth contention at the nearby AP regardless of QoS services supported. Peer applications with the same QoS level will compete for the fixed bandwidth of a single AP, losing any benefit of a QoS scheme. Load balancing will cause some of the peer application traffic to be routed to other nearby APs. Even if a particular AP is close, if it is congested, the collective application user set is better served by spreading the load across multiple APs (also implies good overlapping coverage). This is handled in a proprietary manner in some commercial offerings but will be facilitated by the clients through the network when new 802.11 standards have been ratified and implemented.

6.3.3 Overlapping Coverage Considerations

Ensuring an optimum coverage overlap is not always easily achieved. With 802.11 b and g, channel overlap is a critical consideration when deploying a WLAN for voice.

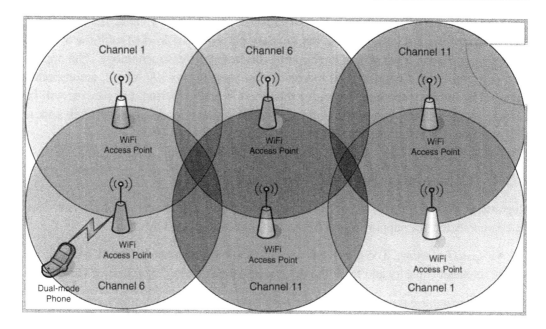

Figure 6.2: 802.11b/g overlapping coverage.

In North America, the Federal Communications Commission (FCC) limits the b/g channels that can be used to channels #1, #6, and #11 (see Figure 6.2). Of the 14 available channels, utilizing these few channels assures a minimum of sideband adjacent channel signal interference. However, having only three channels from which to choose makes mapping out an office environment complex and much like a four-color map problem that challenges many schoolchildren as a standard homework problem. In the same manner that assigning a color to each state without having two adjoining states be the same color is virtually impossible, developing a deployment plan that eliminates any co-channel interference is difficult.

Adjusting the transmit power of the access points is one method used to minimize channel interference, but this option conflicts with the requirement of having overlapping coverage to ensure the best voice experience. Other challenges in optimizing coverage in any one facility are the fact that the very geometry and structural composition of the building will create "nulls" and areas of attenuated interference. Reflective or RF opaque surfaces can block RF radiation and result in coverage "shadows" (weak or no signal) and can also cause reflected radiation to cancel peer signals and result in a "null" location or "dead" spot within a facility.

Because of its 2.4 GHz radio design, 802.11b/g deployments are vulnerable to other sources of interference. The 2.4–2.5 GHz range is an "unlicensed" sector of the spectrum and is labeled the Industrial-Scientific-Medical band (ISM), which is spectrum set aside for commercial applications such as WiFi. Since this is unlicensed spectrum, any company can produce and sell a product that also uses the ISM band for transmitting/ receiving data. Most notable examples are microwave ovens, cordless phones, RFID tags, and Bluetooth devices. Microwave ovens are shielded because of their power levels and are not usually an interference problem, but cordless phones and Bluetooth devices can be a source of interference.[5] It is important that the WLAN be deployed at a density proper for support of voice, and identification of any possible interference source is essential.

Deploying 802.11a does not have such a critical consideration for planning overlapping channel interference, because there are 22 channels from which to select. 802.11a would appear to be the optimum wireless voice technology, but it has not been embraced by the handset manufacturers; only a few mobile phones have been brought to market. Evolution in 802.11a chip design has addressed the power consumption problem, and newer offerings provide 10-hour talk time with up to 100 hours of standby time. 802.11a does have a unique challenge because of its 5 GHz operating frequency range. Due to possible interference with radar in the 5 GHz band, 802.11h was created to define the way WiFi 802.11a devices work with RF mutually coverage problems.

6.3.3.1 Quality-of-Service Considerations

Quality of service (QoS) addresses the traffic priority problem. Voice traffic must arrive at its destination on 20–40 millisecond boundaries. There is no such application performance requirement for file transfers, Web browsers, or general-purpose Internet-centric data applications. Except for extreme network delays (more than 3–4 seconds), these applications function in a user-acceptable manner. For a real-time voice application, however, there is little tolerance for such delays, and unmanaged competition for bandwidth has a destructive effect on any real-time application (voice or video). From a WiFi perspective, the QoS selection requirement would be to identify a mobile handset/PDA that supported the WMM or 802.11e standard. This standard defines the way a mobile device and its associated access point can manage the link between them and give priority to time-critical voice applications over data-based applications. In making such equipment selections, it is important to ensure that the WiFi infrastructure also supports the same level of QoS service as the handset.

---

[5] 802.15.2.2003 is a specification that defines how Bluetooth coexists with other 2.4 GHz technologies.

Beyond ensuring the QoS of the WLAN link, equal attention must be paid to the end-to-end QoS as the network path for the voice traffic traverses routers and the Internet at large. This is often a large, unchartered deployment requirement.

6.3.3.2 WLAN Topology Considerations

Initially, when wireless LANs were introduced to the market, they were implemented as access points (AP)—Ethernet nodes that acted as base stations for mobile devices. These APs were attached to a hosting Ethernet at strategic places within a facility to provide wireless access at the point of need (see Figure 6.3).

Wireless LAN-Enabled Corporate Ethernet

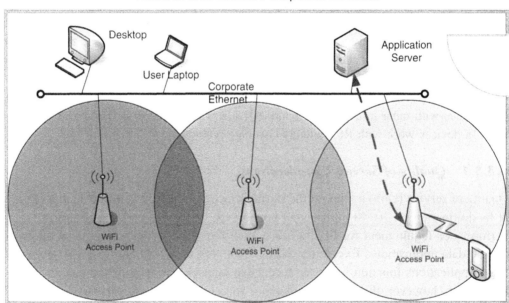

Figure 6.3: WLAN with thick access points.

When a mobile device is attached to the network through a nearby access point, it functions exactly like any other node that is attached through the physical Ethernet cable. Roaming between APs simply changes the data point of entry onto the network. The example in Figure 6.4 shows the data path of an application running on a mobile handset to a target application server. This original topology has been termed the "thick" (or fat) access point model. This is because the AP had all the intelligence built

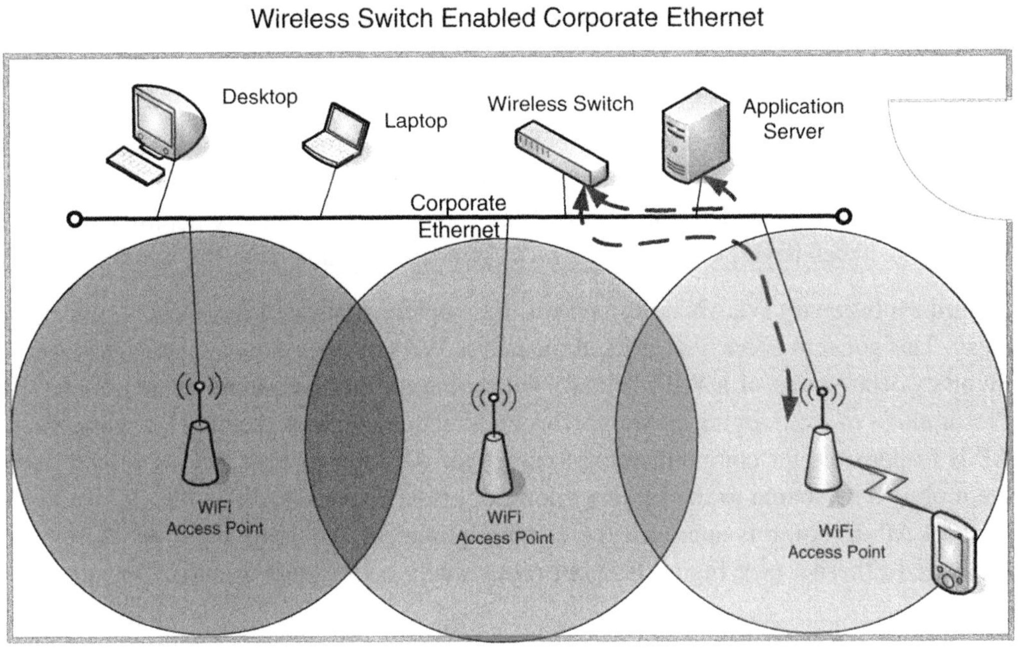

Wireless Switch Enabled Corporate Ethernet

Figure 6.4: Thin AP Ethernet.

in that was necessary to manage multiple mobile devices moving data on and off the Ethernet. Such architecture also resulted in a design where new APs could be deployed independent of any previous configuration but also restricted the WLAN services to a single subnet and required a more costly product, which inhibited many WLAN buy decisions.

In 2002, a new "switched" architecture was introduced to the market that radically changed the WLAN market almost overnight. This architecture used the concept of "thin" (nonintelligent) APs that collaborated with a network resident wireless switch for support of the LAN connection. The distinct advantage of this approach was that the overall WLAN cost burden could be shifted to the switch controller, which resulted in a lower overall solution cost. Figure 6.4 shows that the resulting data path of the same application now goes from the mobile device through the wireless switch and to the destination application server. This circuitous route, though adding some incremental latency, affords new features not possible in the old, fragmented WLAN architecture. Because all the "thin" APs communicate directly with the switch, they can be managed in a centralized manner not available before. Additionally, certain inter-AP operations and client roaming decisions that would have been virtually impossible with a "thick" AP design can now be optimized.

In the current market, all major WLAN vendors offer some variant of a switched network. Each may have its different flavor of security, QoS, or management to provide product differentiation. The end result is that for in-building WLAN deployment, wireless switches have become the *de facto* standard and dominate the market.

A third evolutionary WLAN milestone was the implementation of the concept of a WiFi *mesh*. This concept is one that is ideal for public WiFi or open-area coverage because it involves deployment of a WiFi "cloud" with minimal Ethernet points of presence. Presumptive of overlapping coverage, this architecture employs the concept where each AP is responsible for communicating to neighbor APs and routing its data to and from the mobile data source to the Ethernet point of presence. Employing a "thick" (or even thicker) AP design, this approach forces no requirement for deploying a parallel hardwired Ethernet (see Figure 6.5) but relies solely on wireless transport for mesh

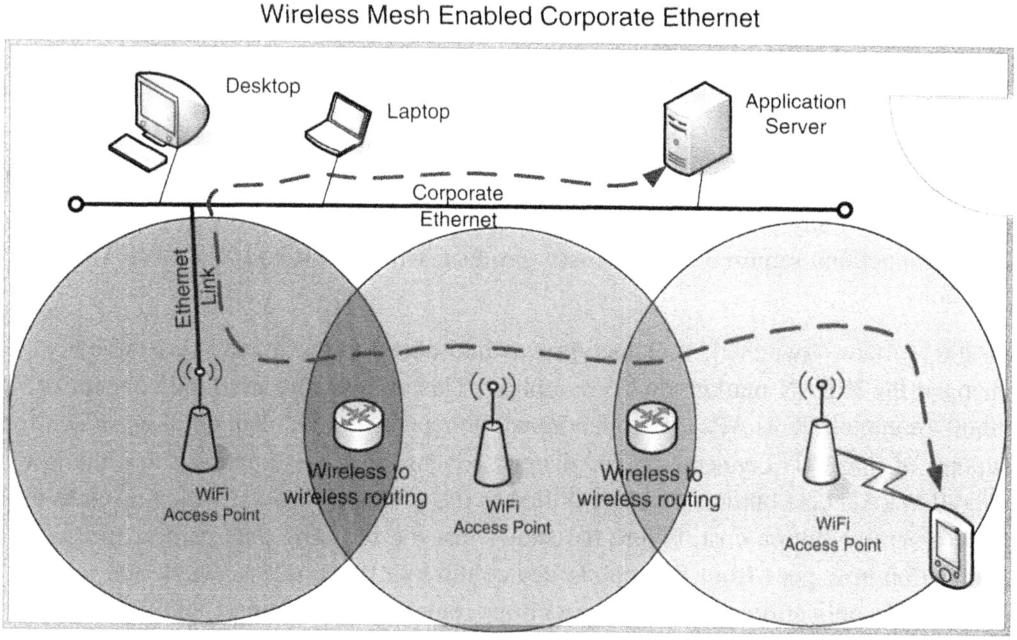

Wireless Mesh Enabled Corporate Ethernet

Figure 6.5: Mesh topology.

operation. To make it commercially viable, each *mesh cloud* must be connected to an Ethernet subnet or to the Internet via a single connection point.

In a "mesh" environment, the data path required to run the example application is now routed through each AP within the "mesh" cloud and then to the application server. This is a topology that can be deployed with minimum network connections but is typically not optimized for real-time applications such as UMC and may add undesirable latency, affecting voice quality. Mesh, however, is often the WiFi technology of choice for public/municipal hotspots.

6.3.4   802.11/WiFi Standards Overview and Status

In an effort to address all the evolving market requirements for wireless LANs, the 802.11 working group spawned a number of *alpha-character* subgroups to address specific requirements within the standard (see Table 6.1).

Voice and other real-time applications can be supported across the existing WiFi standards but are often supplemented by proprietary mechanisms. Final ratification and implementation of the remaining proposed standards will "plug the holes" and provide a solid base on which to build wireless real-time applications.

6.3.5   802.11e/WMM: Quality of Service

802.11e is the Quality-of-Service (QoS) amendment. In the original 802.11 standard, all applications had equal access to the media and would compete for "airtime" to transmit or receive data; thus when a file transfer application was running at the same time a voice application was running, these two applications would compete for the commonly available bandwidth. Because voice is a real-time application, latency and jitter (randomness of packet arrival) have a big impact on voice quality, but with a file transfer, it doesn't matter if the delay is 500 or 3,000 milliseconds.

802.11e was defined to prescribe how applications negotiate for "noncontended" bandwidth when needed for application requirements. Priority to the WLAN media would be given for applications that specified that need, and others would be relegated to a "best-effort" transmit/receive basis. The specifics and options defined in the standard will not be discussed here, but it is important to understand that

Table 6.1: IEEE 802.11 task group charters

802.11 Amendment	Ratified	Description
802.11**a**	Yes	5 GHz high data rate
802.11**b**	Yes	2.4 GHz 11Mbps data rate
802.11**d**	Yes	International RF regulatory conformance
802.11**e**	Yes	Quality of Service (QoS)
802.11**g**	Yes	2.4 GHz high data rate
802.11**h**	Yes	Radar avoidance standard (5 GHz band)
802.11**i**	Yes	WiFi Protected Access: security
802.11**j**	Yes	Japanese-specific RF standard
802.11**k**	Yes	Neighborhood reporting
802.11**n**	Yes	100+ Mbps data rate, final ratification 2009
802.11**p**	April 2009	Vehicular environment
802.11**r**	Yes	Fast roaming
802.11**s**	No	Mesh network architecture
802.11**u**	No	Access to external networks
802.11**v**	No	Client control specification
802.11**w**	Yes	Extended Security specification for management frames
802.11**z**	No	Extensions to Direct Link Setup (DLS)
802.11**aa**	No	Optimization of video applications

to fully support any QoS, both mobile device and WLAN infrastructure products must support the same level of the standard. To ensure and validate conformance, the WFA defined the WiFi Multimedia (WMM)[6] interoperability certification. In making a decision for UMC purchases, it will be important to look for WMM support for both handset and wireless infrastructure, since the "marriage" is vital to good voice quality.

---

[6] Also known as Wireless Multimedia Extensions (WME).

6.3.6   802.11i/WPA/WPA2/WPS: Security

802.11i is the security amendment. The original Wired Equivalent Privacy (WEP) wireless encryption was weak and could easily be broken. Until the IEEE proposed a stronger security standard, the market was stunted and would not grow beyond the early adopters and tire-kickers. The 802.11i amendment specified a more advanced encryption schema (in fact, the Advanced Encryption Standard, or AES, was adopted) and included provision of authentication and authorization.

For adoption by businesses, it was vital that a strong, virtually unbreakable encryption be implemented and some method of authentication be supported. Paired with encryption, authentication is a process by which the two parties recognize each other as being permitted to exchange certain information. In many general applications, this involves a username and password validation. With a WLAN implementation it is important that the handset be validated to the wireless infrastructure and the wireless infrastructure (or application) be validated to the mobile handset; this is the authentication step. This can be accomplished via many methods, and some flexibility is permitted in the 802.11i amendment, but many systems use the 802.1X as an authentication protocol framework. There is some confusion about 802.1X, but it is merely a standard framework or protocol for implementing an authentication schema and not the authentication process itself. Most often the authentication is accomplished by interfacing with a RADIUS server to validate the user or device. Once this has been accomplished, there is a well-defined method for encryption exchange or configuration, and the data between the handset and the AP are encrypted. More details are found in the security chapter.

In like manner with 802.11e, the WFA defined an interoperability certification for security, called Wi-Fi Protected Access (WPA). The original WPA, a subset of the 802.11i standard, was replaced by WPA2, which has stronger security specifications derived from the 802.11i standard. Various security options are supported under WPA, from full negotiation of security keys to WPA-PSK (private shared key). WPA-PSK is often selected for wireless voice deployment due to minimum impact on the voice quality (because of latency) and because it still has a strong encryption method. WPA2 also contains the concept of *preauthentication*. This function allows a mobile device to complete an authentication/authorization exchange with an AP without being associated with it. Such a feature is very important for UMC applications because it significantly reduces the elapsed AP-to-AP roam times.

Setting up an enterprise secure wireless LAN may require detailed planning and knowledge of network details, but to be successful in a home (or nonbusiness) environment, setting up security must be simple. To this end, the WFA created Wi-Fi Protected Setup (WPS). This design requires little networking knowledge and is invoked through simple user operations. The design doesn't scale, but home or SOHO installations are typically small, and WPS greatly simplifies adding a new wireless device to a network. Knowledge and planning for WPS may be required by enterprises in support of any telecommuters or remote office access.

6.3.7   802.11k: Neighborhood Report (and More)

802.11k is the amendment to the base standard that defines how traffic may be distributed in a WLAN environment. The protocol defines a method by which the infrastructure gathers information about the overall WLAN configuration and, when requested, reports to the mobile handsets via a "neighborhood" report that identifies the APs near to the currently associated one for that device. With this information, it will then be possible to achieve load balancing within a network. When congestion is discovered (by the mobile device), the 802.11k information will allow it to make an intelligent decision on roaming off the congested AP.

This standard was recently ratified, but no commercial support is currently on the market. Even with ratification and publishing of 802.11k, there will be implementation issues to consider. In the 802.11k environment, all the mobile stations will receive the *neighborhood* report and will be capable of making a roam decision to load-balance traffic. If, by chance, all devices make an equivalent roam decision and roam to the same AP, we have a "lemming"[7] effect, and the same problem exists but in a different network location. To affect true load balancing, some hysteresis (statistical history) must be retained, and each mobile station makes a roam decision based on a congestion-polling frequency. Such sophistication may only be supported in single-vendor implementations (infrastructure and handset).

---

[7] The *en masse* lemming rush off a cliff was shown by Walt Disney in a late 1950s nature TV show. The entire herd would act as one and rush synchronously over a cliff into the sea. This "fact" was later proven to be a theatrical contrivance purely for ratings enhancement.

6.3.8   802.11u: Access to External Networks

802.11u is the working group that defines the specifics of the proposed Internet work access standard. The intent of this amendment is to simplify access of external 802.11 networks (away from home or office). As more municipal and commercial hotspots are deployed, the need to facilitate seamless roaming across these networks is important. Current IEEE estimates put ratification of this standard in early 2009.

Preceding any standards work, other offerings on the market perform these intended functions. The major hotspot Wireless Internet Service Providers (WISPs) have announced support of such capability and are working with the major handset vendors to integrate this function. Implementation of a seamless (e.g., no user action required) roam from a cellular network to a WiFi hotspot is critical for support of UMC products and will be commercially available in early 2008.

6.3.9   802.11v: Mobile Client Management

802.11v is complementary to 802.11k. Where 802.11k provided information for a mobile device to make decisions, 802.11v will allow some network intelligence to dynamically manage a mobile device's configuration and state—more of a cellular management model. The proposed standard is in the balloting phase and promises to strengthen the ability to manage mobile devices that may be attached to a network. Support of this proposed standard (estimated ratification schedule is September 2009) is not essential for a UMC solution but could better facilitate deployment and management of such a dispersed population.

6.3.10   802.11r: Fast AP-to-AP Roam

The 802.11r task group is focused on defining how ultra-fast AP-to-AP (inter BSS) roams may be achieved. In a real-time application environment (voice or video), traditional nonoptimized roams between APs may take as long as 3 to 5 seconds and would include network scanning, authentication/authorization, negotiating QoS, and establishing security links. WPA2 describes how preauthentication might be achieved, but the actual disassociation and reassociation between two APs still require too much time. The exact architecture specified by this standard has not been finalized, but the goal is to accomplish roams in tens of milliseconds, well under any time constraints for a voice application.

This critical UMC standard is projected to be ratified in early 2008, with product supporting it market ready sometime in mid- to late 2008. Until that time, there are several proprietary methods employed by specific vendors that make claims for speeding up AP-to-AP roams.

6.3.11   802.11s: Mesh Network Standard

The 802.11s TG is still meeting and refining this amendment that is chartered to define how multiple WiFi entities may dynamically create an *ad hoc*, self-configuring multihop topology. In principle, the idea is to create a WiFi "cloud" where adjacent APs set up routing paths through which mobile handsets may have access to a defined network (e.g., the Internet). The "mesh" would be made up of collaborating "mesh points" that could provide:

- Gateway services to a designated network (through a "mesh portal" via a 802.3 connection)

- Proxy authentication/authorization services

- Deterministic routing paths from mesh entry to 802.3 connection point

There is no announced ratification date for this standard, but there are already a number of WiFi Mesh products on the market. They're proprietary at this point, but the assumption is that when the standard is ratified, most (or all) of them will be upgraded to support 802.11s. At this point, the WFA will develop an interoperability certification that will be important for intervendor product integration.

6.3.12   802.11h: Radar and Satellite Interference Mitigation

This standard affects only the 802.11a implementations that utilize the 5.2 GHz bands. In Europe, the government/military radar systems also utilize this frequency, so to prevent interference with these systems, a method for avoiding interference was necessary. Additionally, it was identified that potential interference with satellite communication was also possible. This standard dictates how an 802.11a device or infrastructure must minimize interference through dynamic frequency selection (DFS) and transmit power control (TPC).

6.3.13 The Missing Standards

For the curious, you may note that there are a number of alpha-label standard groups that were not listed in this section. The reason for this is that they are (1) no longer part of the standard; (2) not applicable to the UMC solution; or (3) will not be used to avoid possible confusion; that is, there is no 802.11o, 802.11l, or 802.11x standard to be defined. A status of all these 802.11 task group efforts may be found at http://grouper. ieee.org/groups/802/11/QuickGuide_IEEE_802_WG_and_Activities.htm.

6.3.14 802.11n

The ever-changing 802.11 landscape now has a new wireless technology choice on the horizon: 802.11n (estimated standard ratification, June 2009). Even the 802.11g theoretical data rate of 54 Mbps is seen as insufficient for more sophisticated applications such as high-density wireless video—thus the requirement for a new wireless standard. The benefits of 802.11n are significantly higher data rates, up to several hundred Mbps, and improved range and coverage. The higher data rates are achieved through implementation of several parallel transmission techniques:

- *Multiple-input/multiple-output (MIMO)*. This is achieved through use of multiple antennae that can be used concurrently to transmit/receive data.

- *Channel bonding*. The technique of parallel transmission in multiple nonoverlapping RF channels.

Additional bandwidth enhancements have been made through *packet aggregation*. This is a more effective structure for data packing of each packet to maximize the data transmitted on a per-packet basis.

The market demand for higher wireless data rates is so strong that the WFA launched a program to certify 802.11n-Draft2 products even before the standard was ratified. Such prestandard products are now in the market, with the promise of upgrading to the final standard when ratified. Current projections of final ratification will be sometime in late 2008 or early 2009.

Will UMC solutions be able to take advantage of 802.11n? There are several technical challenges that need to be overcome before support can be realized. Effective RF design challenges are faced with attempts to support MIMO on a handheld device; antenna design is very complex and complicated by the handheld form factor. Additionally, increased power demand will pose a battery life problem that will need to be resolved before products can become commercially available. The bigger question must be asked as to whether handheld devwill ices really need the bandwidth offered by an 802.11n network? Smaller screens mean less data to be transmitted, even for video applications; 802.11g or 802.11a works just fine for such applications.

Though 802.11n holds great promise for prominence of wireless LANs as the backbone network, 802.11g and 802.11a fill a demand that will remain for a number of years to come.

Beyond 802.11n, the IEEE is looking at the following wireless generation that has been labeled Very High Throughput (VHT) and that may operate in frequencies below 6 GHz and at 60 GHz.

6.3.15   802.21

Begun in 2004, this working group was chartered to define a standard for media-independent handover (MIH). That is, where devices equipped with the proper network connectivity could perform seamless roaming between two disparate networks (WiFi, Bluetooth, WiMAX, Ethernet, and/or cellular). The standard doesn't define how the handover is accomplished but rather provides layer 2 reporting and services to facilitate such a handover. The current projected timeline for ratification is 2009–2010.

6.4   IEEE 802.16/WiMAX: The Future Looming

One newly emerging technology from the IEEE is 802.16, better known as WiMAX. This marketing acronym was assigned to this standard by the WiMAX Forum and stands for Worldwide Interoperability for Microwave Access. The WiMAX genesis was a need for a last-mile wireless connectivity solution that would compete with residential/business cable and DSL services and could front end for wireless or wired national network infrastructures (see Figure 6.6).

There has been a lot of press and industry chatter about WiMAX as well as some misconceptions relating to this proposed IEEE standard. Public references to "fixed" (802.16d) and "mobile" (802.16e) WiMAX have blurred the actual state and content of the standard. In truth there is no IEEE 802.16d or 802.16e but rather ratified 802.16-2004 and 802.16e-2005. The 2004 version of the 802.16 specified what is being termed *fixed* WiMAX, an architecture without base-station-to-base-station roaming. The 2005

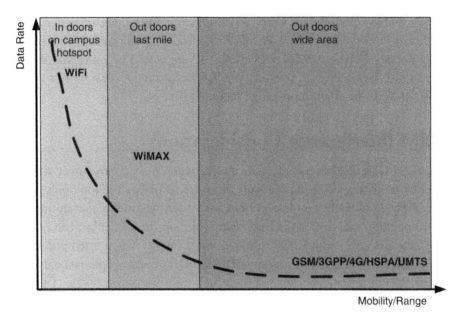

Figure 6.6: WiMAX market positioning.

version is often termed the *mobile* WiMAX and is more like WiFi in its ability to support devices roaming between base stations within a WiMAX network; it has significant enhancements defined in the encoding and antenna design aspects to optimize for bandwidth in each media.

Initial WiMAX offerings will be built on the 802.16-2004 specification, with 802.16e-2005 products coming to market in late 2008. Prestandard offerings in the form of WiBRO (sponsored by the Korean government) and Xohm (Sprint's expression of WiMAX) have already been announced. WiMAX is unlike WiFi in that it is not limited to the unlicensed ISM spectrum but can also be implemented in a licensed spectrum. The current perspective is that WiMAX will not cannibalize the WiFi market but may erode the wide area wireless networks for municipal and commercial last-mile hotzone support. Some carriers have also shown strong interest in utilizing VoIP over a WIMAX fabric to target direct cellular market subscribers (WIMAX handsets) or for use in cellular backhaul technologies.

How will WiMAX play into the UMC market? That answer is yet to be addressed. Conceptually, a *trimode* phone is possible (WiFi/mobile-WiMAX/cellular), but such technically possible solutions are years away, pending addressing of battery, form factor, and cost issues. More likely WiMAX will make its appearance in municipal or community hotspots where a WiMAX/cellular device would be more applicable. WiMAX can also be used in an overall wireless infrastructure design as a backhaul technology for WiFi or other short-range technologies.

6.5   ISM Interference Considerations

Because the 2.4 GHz ISM bands are unlicensed, there is no control over what devices may be deployed in any one site that uses these frequencies. Technologies such as Bluetooth, Zigbee, 2.4 GHz cordless phones, or RFID tags may generate interfering signaling competing with any 802.11b/g UMC solution. In particular, use of Bluetooth hands-free headsets can result in deteriorated voice quality when attempting calls over WiFi due to the close personal proximity. The 2.4 GHz cordless phones can also cause considerable RF interference because of their high transmit power levels (1 watt vs. 100 milliwatt). Due to these possible problems, any UMC installation should include an interference assessment before going into production mode.

CHAPTER 7

VoIP: Layers 3 and 4, the IP Infrastructure

7.1 VoIP: An Introduction

From the time Alexander Graham Bell patented the telephone through the latter part of the 20th century, all telephony products were built around a circuit-switched design. Whether analog or digital, this meant that to set up a call between two points, a physical metallic circuit had to be established and was dedicated for the length of the call. Because such technology has matured over the past 100 years, we have become accustomed to the reliability of this infrastructure; we know that when we pick up a receiver (going off hook), we'll get a dial tone.

Inherent in such an architecture was an upper capacity limit of the number of simultaneous circuits that could be sustained. Because the profit margins were so high in the telephony industry, it was not a major problem to continually add more discrete circuit elements to support the ever-growing need for capacity. There was, however, one major flaw with the circuit-switched approach: There was no simple way to integrate other forms of data communication, such as data and video. Extensions to telephony standards were devised to support transmitting fax and data over the PSTN, but these were all serial operations and stretched the circuit-switched design beyond its original concepts of supporting just voice. The basic telephony solution remained an isolated service, and other communication technologies (video and data networking) developed independently and were nonintersecting. There was a great gulf between voice and data/video communications technologies.

In the last 15 years of the 20th century, a disruptive voice transport concept was born: voice over a packetized (noncircuit-switched) network. This is a radical change and involved taking digital samples of voice, packetizing them, and transmitting them on a network highway to an end point that would reassemble the audio streams and replay them. Such a highway, the Internet, had come into existence, and Voice over IP (VoIP) was seriously considered as a new option for transmitting voice from one point to another.

The thought of packetizing voice and sending it over an IP network now offered many benefits not realized with the circuit-switched infrastructures. Video and application data could share this highway with voice, unifying the network requirements and greatly simplifying total cost of ownership (TCO). Additionally, multiple peer-to-peer calls could be accommodated through interleaving the packets as they traversed the Internet; such a situation allowed for *self-healing* of routes through the network if one component failed, which was not the case with circuit-switched designs. Very quickly academics, telephony vendors, and standards bodies saw the exciting opportunity of VoIP, and the race was on.

Over the next few years, multiple initiatives were launched to design and develop viable VoIP solutions that could support all the functionality of the legacy circuit-switched networks and extend the functionality beyond that base converging voice, video, and data. As with most new business and technological opportunities, the first solutions to arrive on the market were proprietary. These were eventually followed by multiple international standard VoIP solutions. As with commercialization of any technology, multiple VoIP options come to the market (like Beta-Max and VHS in the earlier video market), and as with video, the market eventually selects a single expression of that technology that will dominate.

7.2   Voice-over-IP Protocols

The following sections review some historic VoIP protocols and provide an overview of the dominant standard, Session Initiation Protocol (SIP), now supported by almost all players in the VoIP market.

7.2.1   VoIP Protocol Soup

A VoIP protocol is an OSI layer 4 protocol (above layer 3, TCP/IP) that manages call setup, call monitoring, and call termination. Vendor proprietary protocols were the first to market almost 10 years ago due to the fact that the standards didn't exist or didn't address all the functional requirements necessary to support an IP handset. These extended features include facilities for controlling the phone display and keypad for management of specific functions accessible by the user. Some of the most popular early vendor VoIP solutions were:

- *Cisco's Skinny Client Control Protocol (SCCP).* Base control protocol for Cisco's Unified Communication Manager VoIP solution.

- *Nortel's UniStim.* Base protocol used in support of Nortel's 2004 IP phone solutions.

- *Mitel's MiNET.* Base protocol used in support of Mitel's ICP IP PBX solutions.

- *Alcatel's UA.* Base protocol used in Alcatel's IP PBX solutions.

- *3Com's H3.* Base protocol used in 3Com's NBX IP PBX solutions.

Many of these protocols may be found in operational deployments even today. However, as the VoIP market matures, there is a definite shift to support of standards-based products.

In parallel, several standards bodies worked to create international standards that would guarantee reliable voice with intervendor interoperability. The most popular international standard protocols are:

- *IETF Session Initiation Protocol (SIP).* SIP is the dominant VoIP standard, being adopted by virtually all VoIP vendors (RFC 3261).

- *ITU H.323.* The early VoIP leader, but it had a complex design.

- *ITU H.248/Megaco.* Another international standard defined and implemented by a number of vendors.

Besides their proprietary natures and incompatibility, one design difference between most vendor proprietary solutions and the international VoIP standards was the basic architectural paradigm. Many vendor designs followed a "dumb client/smart server" model, which has been labeled a *stimulus design*. This is because the server (i.e., iPBX) sends commands to the handset, where there is no intelligence or state information, and the terminal simply responds. Such an approach allowed vendors to expand their stimulus design beyond simple support of basic telephony commands to include services to drive features on the target handset, which were outside the scope of most VoIP standards. The International VoIP standards tended to follow a "smart client/smart server" (or true client/server) model. This design approach allows creation of products from different vendors that interoperate, will accelerate the market, and are essential for any UMC implementation.

7.2.2   Stimulus Protocols

The legacy of the TDM telephony industry was that the solutions were very proprietary and one vendor provided both PBX and handsets/desksets; there was no option for intervendor interoperability or competitive aftermarket solutions. This fact allowed the margins to remain high for these companies, protecting their market positions. This proprietary concept was propagated to the emerging VoIP solutions, and because a single vendor had control over both ends of the product set, it designed protocols that were hardware specific (e.g., turn on a light, write a string to the third line on the LCD, or even display stock quotes on the LCD). In such a manner, one protocol could be developed that evolved in complexity as the deskset/handset form factors evolved; the intelligence of the application remained within the core of the vendor's PBX product.

7.2.3   Client/Server Protocols

With a client/server VoIP model, the intelligence for the total application was partitioned between the two participating elements: the deskset and iPBX. Such a model could have been adopted by the legacy PBX vendors, and the result would have been a proprietary client/server solution; this was not generally the case.[1] The client/server model was selected by the international standards bodies because it offered an opportunity for multiple vendors to develop and market the different elements in a VoIP solution; it leveled the playing field for competitors because of the guarantee of interoperability. Table 7.1 details the pro and con features from the perspective of the VoIP provider.

The reason the legacy PBX vendors chose a *stimulus* approach is clear from Table 7.1, but the tide of market adoption has caused a shift to force all UMC vendors to support the client/server model because of a guarantee of intervendor interoperability.

---

[1] There were a few standards-based, vendor-specific implementations that confused the market. The claim was to support the standard but without guaranteed interoperability with other vendor products that also conformed to that standard.

Table 7.1: Stimulus vs. client/server VoIP comparison

VoIP protocol model	Pro	Con
Stimulus protocol	• Allowed support of hardware-unique control features • No extravendor interdependencies (single vendor solution) • No limits on the deskset functionality; could go beyond just telephony features • Blocked out potential deskset competitors from the market	• Required changing the protocol with each new deskset form factor or feature enhancement
Client/server protocol (international standard defined)	• Leverage off the market adoption of a specific standard—a marketing advantage and not a technical advantage	• No hardware-unique control features—limited to just telephony or multimedia-supported features[2]

7.2.4   VoIP Feature Requirements and Concepts

For all client/server VoIP implementations, some common concepts and feature requirements are necessary to support a robust application framework.

7.2.4.1   Voice Encoding

In commercial phones, the analog voice is converted into a digital form by applying what is known as a coder-decoder, or *codec*. A codec is an algorithm by which an audio sample is encoded into a binary object that can be transmitted and replayed via a reverse operation. There are multiple codecs standards that are used in the telephony and cellular industry, where each codec has its own unique characteristics that make it more applicable for a specific purpose (see Table 7.2). There are too many codecs to discuss thoroughly in this book, but suffice it to say that any two end points exchanging audio information must support a mutual codec. The codec is usually negotiated when a call is established and is the lowest common denominator between the two call participants.

---

[2] Support of extended features had to be implemented "out of band" outside the scope of the VoIP protocol definition.

Table 7.2: Common codecs

Codec	Size	Comments
G.711	64 kbit/s	Baseline codec supported by all telephony applications
G.729	8 Kbit/s	Most popular codec but requires a royalty fee for the patent holder
G.726	16 Kbit/s (and greater)	Also known as Adaptive Differential Pulse Code Modulation (ADPCM); now obsolete, the standard was a popular codec for some solutions
G.723	24 and 40 Kbit/s	Common codec used in IP desksets
ILBC	15 Kbit/s	Popular GSM codec used in some VoIP applications

7.2.4.2 Packet Size/Rate

The quality of a VoIP call can be largely affected by end-to-end latency and jitter. In most implementations, audio is transmitted on 20- to 30-millisecond boundaries to afford a steady stream for replay. Such an implementation, however, cannot guarantee good voice quality without attention to other configuration factors.

Total end-to-end latency is important for satisfying the UMC user. If the end-to-end latency is greater than 400 to 500 milliseconds, the conversation between two individuals becomes clumsy where they overstep each other. This latency is due to any number of network conditions:

- *Traffic congestion.* Occurs if one of the subnet trunks that are used is congested with general network traffic without QoS or bandwidth management.

- *Packet loss and retransmission.* Any protocol retransmissions will incrementally add to the experienced latency and incrementally impact voice quality.

- *Packet out of sequence.* In a packet network the packets could arrive out of transmit order. This requires the receiving end to reorder the packets prior to replay.

Jitter is a way of expressing the disparity between packet arrival times. If the arrival time between two packets is greater than the expected packet sequence time, the late packets are discarded, which results in voice quality deterioration. Jitter buffer management may be viewed as a "shock absorber" for audio traffic that smoothes out the bumps in packet arrival times, better guaranteeing improved voice quality. There will always be some jitter, but the frequency and magnitude of the jitter can severely impact the user experience.

7.2.4.3 Quality of Service

Because VoIP is a real-time application, it is vital that a steady stream of audio packets arrive successfully (and on time) for replay. In a general Carrier Sense Multiple Detection/Collision Detection (CSMD/CD) architecture such as Ethernet, without some *service-level* guarantees, all packets have the same transmission priority. This means that a voice packet stream may have to compete with packets associated with a file download for access to the LAN media. The file download is not impacted if it takes an additional 2 to 3 seconds to transmit some frames, but the voice application is severely impacted with such conditions. It is important, in a multiuser LAN environment, that some QoS guarantee be provided.

Implementing and supporting the required QoS for UMC applications, however, is very complex since the connection may traverse many different subnets, network topologies, and management domains. Management of a QoS service becomes a multitiered challenge, with priority being implemented at the WiFi, wired network, and intranet levels. For enterprises, management of the first two domains is achievable through the network management team. For consumers, there is virtually no guarantee of priority on real-time applications unless they are routed through a VoIP managed network from a service provider.

7.2.4.4 Bandwidth

Each full-duplex (two-way) call requires a fixed bandwidth to support the call along the network path. The selected codec dictates the basic bandwidth required, which for G.711 (64 Kbit/s) can be 140–180 Kbit/s demand for a single call. Aside from QoS considerations, there is typically enough bandwidth on a hardwired network to support multiple calls (four or five calls on a slow 1 Mbit/s WiFi link), but packet prioritization (QoS) and load balancing play an important role in assuring good voice quality. The end points of a VoIP call have no control over the available network bandwidth; therefore, adequate network bandwidth planning and management are important for a successful VoIP deployment.

7.2.4.5 DTMF Support

The utilization of Dual-Tone/Multi-Frequency (DTMF) technology is pervasive. Developed in the 1960s to replace pulse dialing, the new *touchtone* dialing was achieved by simultaneously transmitting dual-tone audio signals that mapped to dialing digits. A pair of dual-tones was assigned to each of the digits or characters (1–0, *, #) on the dialing pad of a phone. To make a call, the phone was placed off-hook and the phone number was dialed.

Support of DTMF becomes critical in a VoIP solution for access to an Interactive Voice Response (IVR) system in using the dialing pad to enter the name of a person to select a phone extension from a corporate directory, enter a number in filling an automated pharmacy prescription, or answer some questions for proper product support call routing. Support for DTMF over VoIP allows the user experience (and all the infrastructure systems that use it) to remain the same. More about DTMF later in this chapter.

7.2.5 VoIP Standards

As indicated at the beginning of this chapter, a number of international VoIP standards have been developed and deployed. Of these, only two have achieved any level of significant market traction, and we will take a look at these in the sections that follow. Of the two successful VoIP standards, however, only one seems to be in place to achieve global adoption.

7.2.5.1 ITU H.323

H.323, brought to market in the mid- to late 1990s, was one of the first VoIP standards to be commercialized. This standard had evolved from a videoconferencing standard to one that could also embrace generic VoIP applications. ITU H.323 is an application-specific protocol in which the basic structure and service definitions are tightly bound to video and audio support. Setting up, managing, and terminating voice and video links define the core of the H.323 benefits. Because of its design legacy, H.323 tended to be complicated for the implementation team but was successfully brought to market by a number of companies, including Microsoft with its original NetMeeting product. The H.323 architecture also defined user authentication/authorization, midcall management features, flexible encoding (codec), and support of PBX supplementary services (transfer, conferencing, park, hold, and the like). The H.323 standard was also first to utilize the International Engineering Task Force (IETF) Real-Time Protocol (RTP, RFC

1889) for transmitting audio streams on a packet network, an audio protocol that is supported by virtually all VoIP products on the market.

H.323's early market success was stunted with the advent of the Session Initiation Protocol specification. SIP was an architecture that was more flexible with regard to support of multiple classes of applications and a better fit for providing services in the evolving IP dominated world. H.323 can still be found in the market, but its overall market presence will eventually evaporate.[3]

7.2.5.2 IETF Session Initiation Protocol (SIP)

The Session Initiation Protocol is less a direct competitor to other VoIP protocols than an abstract protocol framework for setting up multimedia relationships (e.g., "sessions") between two network entities. Within the bounds of the SIP architecture, there is virtually no limit for which these *multimedia sessions* may be purposed; support of VoIP is but one such possible use of SIP. The semantics of SIP are based on a text format and therefore simpler to implement and debug than a more complex protocol.

Definition of the session functionality is negotiated via the Session Description Protocol[4] (SDP), an ancillary standards family that provides a mechanism to dynamically negotiate session parameters for any multimedia session, including:

- Protocol version and owner

- Session name

- Email address, phone number

- Media specification

- Bandwidth requirements

- Time zone

- Encryption key

SIP itself has a rather small set of protocol elements that are used to establish, monitor, modify, and terminate a session. Figure 7.1 is an example of a simple SIP phone call placed between a mobile extension and a desktop extension through the in-house iPBX.

---

[3] http://en.wikipedia.org/wiki/H.323.
[4] RFC 2327 and associated standards specify SDP.

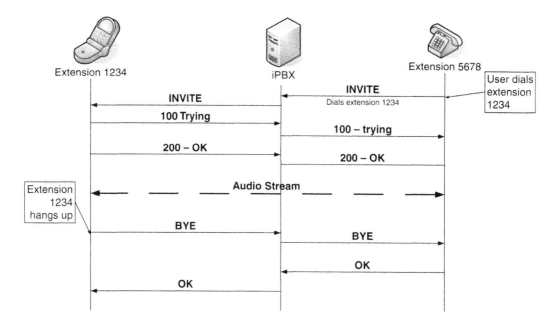

Figure 7.1: A simple SIP example.

Extension #5678 dials extension #1234, which initiates a SIP sequence to establish the session via the INVITE, Trying, and OK sequence. Once established, the audio stream is typically transmitted between the two end points. Termination is by #1234, which sends a BYE command through the iPBX proxy echoed to the other extension. Acknowledgment (ACK) of the BYE completes the session and terminates the audio stream.

SIP can be implemented as a bifurcated architecture whereby (1) signaling and (2) audio processing are decoupled. Such a design allows for variances in implementations that take different expressions with respect to UMC products. One approach employs splitting signaling path from the audio path where an intermediate *proxy* (iPBX or gateway) is involved. In this case, the signaling is passed through all three call components, but the audio stream is directed between the two end devices (see Figure 7.2). Such a design offloads processing overhead from an iPBX but minimizes midcall management and monitoring possibilities.

For premises UMC applications, it is often a better approach to route both signaling and media through a *mobility* controller as a network control point (see Figure 7.3). Though

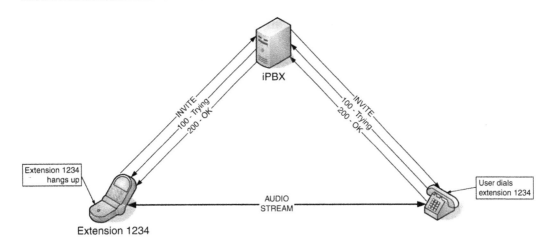

Figure 7.2: Signaling and audio segregation.

this creates the potential for a single point of failure, the fact that both data classes are being managed by a single component allows for a richer management environment that can be more responsive to changing network conditions and minimizes the number of devices necessary for a complete solution. This architecture becomes more important in considering simplification for management of network handovers and support of unified security policies.

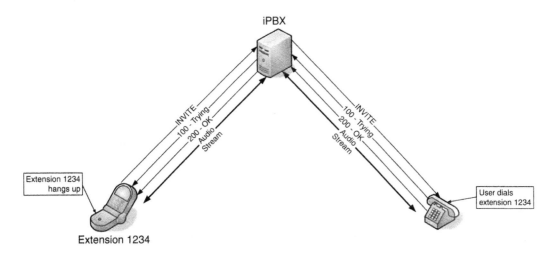

Figure 7.3: Signaling and audio converged.

Table 7.3: SIP requests

Request	Description	RFC number
INVITE	Invites a client to participate in a call session	3261
ACK	Acknowledges a completed INVITE	
BYE	Terminates a session between two entities	
CANCEL	Cancels pending actions prior to establishment of the call	
OPTIONS	Requests capabilities of servers	
REGISTER	Registers the sending User Agent to a SIP server	
PRACK	Provisional acknowledgment	3262
SUBSCRIBE	Subscribes to an event from "notifier"	3265
NOTIFY	Notifies a subscriber of a new event	
PUBLISH	Publishes an event to a SIP server	3903
INFO	Opaque message envelope between two entities in session	2976
REFER	Invokes a call transfer	3515
MESSAGE	Text Message mechanism between two entities	3428
UPDATE	Method for changing the session state without changing the dialog state	3311

The basic SIP requests (see Table 7.3) can be used to create a feature-rich application set that is nonproprietary and permits end customers the freedom of selecting from the best of breed in the competitive open market.

One major challenge for voice protocols in general is how to handle DTMF signaling. DTMF is typically used for accessing corporate employee directories, whereby a user may key in the letters of an employee's name to search for a phone number. In most PBX auto-attendants, to retrieve the phone number for Bob Smith on a non-QWERTY keyboard, you would key in *2266622077776444844*[5] on the numeric phone pad. Each press of a numeric key sends a DTMF tone to the receiving station, which interprets these sequences as letters of the name. Aside from the problem of accurately keying in all these digits, there is often a bigger problem: The DTMF signaling may be corrupted

---

[5] 22 = B, 666 = *o*, 22 = *b*, 0 = space, 7777 = *s*, 6 = *m*, 444 = *I*, 8 = *t*, 44 = *h*.

and not received correctly due to transmission errors. These errors disrupt the name-processing sequence, requiring the user to start the sequence over again.

The most unreliable method for transmitting DTMF is "in band," where the binary audio signals are integrated into the audio stream and appear as injected sound. It is incumbent on the receiving side to detect these signals, apply the proper application logic, and remove them from the audio stream so that the far end user does not hear them. Utilization of existing IETF standards provides a more reliable method of transmitting DTMF, greatly simplifying DTMF processing. RFC 2833 describes how DTMF may be injected into an RTP audio stream not as embedded audio but as discrete data elements that are reliable and easily processed. It is important that any purchase decision of a VoIP solution weigh heavily on its support of RFC 2833.

7.3 Real-Time Protocol (RTP)

Currently, all existing VoIP standards utilize Real-Time Protocol to transmit/receive binary encoded audio samplings. This standard specified the format and options for transmitting audio over either User Datagram Protocol (UDP) or Transport Control Protocol (TCP). A corresponding control protocol, Real-Time Control Protocol (RTCP), is also specified and optionally implemented by UMC vendors. End-to-end secure transmission of RTP frames is specified by the Secure Real-Time Protocol (SRTP). This common transport format simplifies implementation of gateways, proxies, and services, resulting in a relatively high degree of interoperability between audio service points.

7.4 VoIP Everywhere?

The fact that the wide area wireless carriers now provide an IP packet-based service would tend to suggest that you could run a VoIP application over not only a corporate Ethernet LAN but over the wide area wireless network. There is no inherent technical reason that this could not be possible; IP is everywhere, right? However, there are several roadblocks that would impede running a UMC application in IP mode over today's wide area wireless networks.

One technical hurdle is the fact that the 2G and 3G networks don't have sufficient bandwidth or data rate to support a pure VoIP environment of any capacity. But more critical than any technical hurdles are the business hurdles imposed by the carriers due to their concern over the potential of VoIP erosion of their business base. Most certainly, being able to set up telephony circuits over the intranet (bypassing the wireless networks) is a business concern of the carriers, but to use the packet services of the wireless networks for a competing purpose is a real carrier concern, bringing a potential bandwidth demand increase without a corresponding revenue uplift.

Though not universal, some major wireless carriers have implemented network filters to block use of packet services for real-time applications such as VoIP. Some have implemented a timeout on assignment of IP addresses, where the application would have to renew the IP and potentially drop a VoIP call. Some have collaborated with handset vendors to enforce half-duplex packet services, which would allow one to browse the Internet but not establish a full-duplex VoIP phone call. Others go further to implement deep packet inspection and block specific IP traffic types, including UDP and RTP (VoIP audio) frame types.

Given these potential limitations in accessing the appropriate services on wide area networks, how could a successful UMC application be implemented? For the foreseeable future (next two to four years), audio services over the wide area network will be over the standard audio barer channels, like any standard cellular phone call. Supplementary services such as presence and IM would still rely on the carrier packet services, but any audio-related application would not depend on packet services. In this manner, the carriers could not impose any restrictions for use on their network, because such an option would be no different than any other person placing a cellular phone call. With 2G and 2.5G networks there would be some limitation of functionality because the audio and packet services can only be used serially. With 3G networks, access to audio and packet services will be simultaneous.

So, what's the answer to this section heading question: Is VoIP everywhere? The answer is no for UMC applications. VoIP will be the standard/design of choice for LAN and Internet access and carrier bearer channel services for voice. This will be true until such time as the wireless carriers can provide a high-speed packet service with sufficient bandwidth to equal the Internet.

7.5 Commercial Consumer VoIP Services

Millions of Internet users around the world have signed up for Skype, a peer-to-peer VoIP application. Purchased by eBay in 2005, this highly successful VoIP application can be downloaded for free, and phone calls with friends (half the world away) can be placed without charge. At any one time, there may be between 7 and 10 million people globally logged on to Skype and available for a VoIP phone call or IM chat session.

Skype has extended its services to model a traditional central office (CO) from the phone company by providing SkypeIn, SkypeOut, and videoconferencing services. These services support access to the PSTN, where anyone can call or receive calls from anyone else in the world—for a fee.

Skype provides a highly socialized voice application, but it does not implement any of the broadly accepted VoIP standards (such as SIP) and, therefore, provides little integration possibilities to enterprises or into federated associations. Skype has been found to be most useful as an ad hoc phone/IM application that supplements the standard telephony service, supplanting neither desktop landline phone nor mobile phone. It is unlikely that Skype will ever embrace any UMC class services.

Vonage is a successful carrier-class VoIP service provider initially targeted at the home/consumer market. The marketing pitch was that a customer could retain his old analog phone but access the PSTN through the Internet via a Vonage gateway. The company's initial successes were based on cost savings over competing analog/digital home services, but it has now become more competitive by adding peer features and even offerings for the business/enterprise market. Much like Skype, Vonage merely replaces the underlying legacy voice network with a packet-switched voice network and doesn't compete in the PBX or mobility markets. It is unlikely that Vonage will expand its product base to include mobility features, because such product offerings are too far from the company's core competency.

CHAPTER 8

Voice-Optimized Networks: The Network Orphan

8.1 General Network Optimization Considerations

A UMC application often has an unaddressed requirement critical to its overall success: *transport bandwidth availability*. The UMC voice and signaling data may traverse many network segments, depending on locality and service accessibility. Each WiFi, Internet route, and cellular coverage domain may offer varying levels of bandwidth to support the active call. If *any* subnetwork segment has mediocre or suboptimal transport services, the user experience can become unacceptable.

What are the factors that impact network bandwidth availability? Can they be accommodated or avoided? Who controls the critical network segments? These are questions that arise when considering a bandwidth UMC solution, but the answers are often diverse because no one entity controls all possible network paths. UMC can cross public⇔private⇔public network domains, each with unique QoS and bandwidth characteristics under the control of different entities (business or service provider), and may have insufficient or incompatible service levels to sustain the call. In a public, off-campus scenario, it is often assumed that the cellular network will provide sustaining services, but this is not always a fact.

This chapter attempts to address these questions by discussing the factors, solution options, and complications of utilizing a multimanagement network architecture. There is no simple one-size-fits-all solution, but there are solutions that allow a UMC deployment to sail its way through the sometimes rough waters of wireless bandwidth service challenges.

Historically, the issue of bandwidth management has been treated as an *orphan*. For both Ethernet (802.3) and WiFi (802.11), the QoS and bandwidth management elements of the standards emerged some eight or more years after the initial standard ratification.[1] This delay was probably an artifact due to the natural maturing of each of the technologies and the target market demands. The thought of running a voice application over Ethernet was alien in the 1980s, but as data rates increased and product costs plummeted, the possibility of VoIP quickly became a reality. Today, networks supporting the most current IEEE standards (wired and wireless) are well equipped to provide good QoS and bandwidth in support of great voice quality. However, not all commercial networks have provisioned these optimizations, nor have many mobile devices fully supported these key standards. Planning a successful UMC deployment should include understanding the bandwidth limits and optimizations available across all possible network segments.

8.1.1   Network Congestion

Network congestion and transmit collisions have always been problems inherent to networking. In developing the Hawaiian island ALOHAnet in the early 1970s, the concept of collision control was implemented, whereby a time allowed for transmission of data was synchronized by a start-frame indicator. Any node ready to transmit would attempt transmission on this frame mark but would have to detect any collisions caused by another station also attempting a transmission. If a collision was detected, each station would back off a random number of transmit slots before retry; this method of congestion control became known as *Slotted Aloha* (see Figure 8.1).

Slotted ALOHA was not very efficient in managing congestion problems or for bandwidth utilization; more sophisticated methods were applied to the evolving network technologies. In general, network congestion can be handled in one of several ways:

- *Carrier-sense techniques.* Each node will sense the media to see if some other node is "talking" and will back off by some random time period before a retry. Ethernet implemented Carrier Sense Multiple Access with Collision Detection (CSMA/CD) for managing this problem.

---

[1] 802.11e was ratified in 2005, eight years after the 802.11 standard; 802.1p was ratified in 1998, 15 years after 802.3.

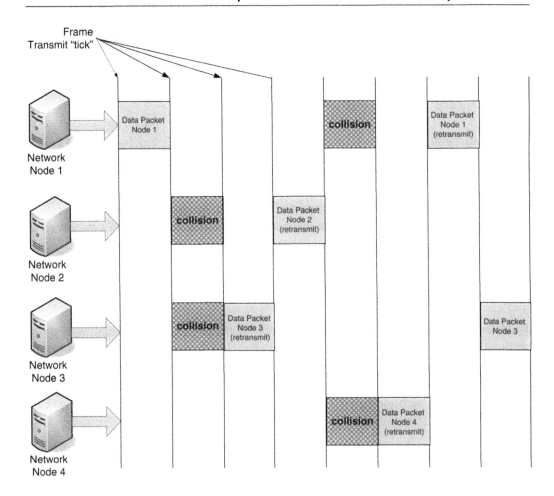

Figure 8.1: Slotted Aloha example.

- *Token-passing techniques.* Each node will be given *permission* to transmit based on possession of a network "token." The early LAN technologies of ARCnet and Token Ring implemented this approach to multiple access control.

- *Time-domain multiplexing.* This scheme assigns each node a *time slot* in which to transmit/receive data. This is typically negotiated when the node attaches to the network and guarantees that there are no collisions, but it can waste unused bandwidth.

These solutions are end-node specific and operate at the physical layer, which is only one aspect of bandwidth management problem with respect to the viability of supporting real-time applications on a LAN. In considering designing a total solution for ensuring application bandwidth availability across the intranet (with multiple subnets), several additional solutions are possible:

- *Shared media access allocation.* This requires reserving bandwidth on a transport media for a specific application to ensure no media contention (resource management).

- *Shared media access prioritization.* This presumes some level of media contention but provides for a mechanism of prioritizing one data type over another in accessing the media (quality of service, or QoS).

8.2   Shared Media Allocation

To ensure optimal VoIP voice quality, it is important that all network segments have minimal congestion levels to support the real-time, low-latency transport requirements of voice. Ensuring that sufficient bandwidth is reserved or managed for these applications can be achieved in several ways. Congestive traffic can be isolated on a managed LAN by partitioning traffic into virtual LANs (VLANs), or a *reservation* can be made between the two participating end points.

8.2.1   VLAN Partitioning

Frames are tagged with a VLAN ID and are routed by switches to specific destinations within a single manage network domain. The prime impetus for a VLAN, isolation of broadcast and multicast traffic, is often utilized to segregate voice from data traffic on a shared LAN network. Figure 8.2 provides an example in which a single wireless/wired network is partitioned into three unique VLANs. By leveraging the VLAN capabilities of the WiFi subsystem to assign multiple ESSIDs on a single access point, wireless traffic from the mobile terminals is isolated and mapped to an equivalent wired LAN VLAN ID. In this way, traffic from mobile *APP1* can never interfere or access services provided by the *APP2* server.

Many modern wireless LAN products and Ethernet switches support the concept of bandwidth management based on assignment to a specific VLAN ID. In this

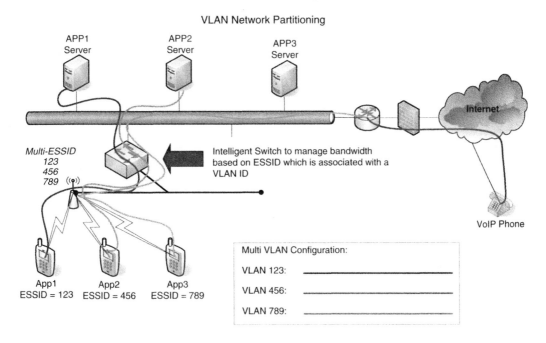

Figure 8.2: VLAN bandwidth management.

manner, the theoretical bandwidth of the combined wireless and wired networks can be partitioned to certain applications that may require uncongested bandwidth.

Local congestion management may be effectively applied via the use of VLANs in headquarters or remote offices where a company IT department has control over the network topology. However, VLANs cannot be applied to reserve of bandwidth for traffic that spans multiple router domains.

8.2.2   ReSerVation Protocol (RSVP)

There is an RFC that can be used for guaranteeing bandwidth across router domains: ReSerVation Protocol (or RSVP; RFC 2205, see Figure 8.3). Designed to address the congestion problems found in a UMC application, RSVP holds the potential of being able to specify a level of QoS for a specific application instance that can be dynamically applied on each call.

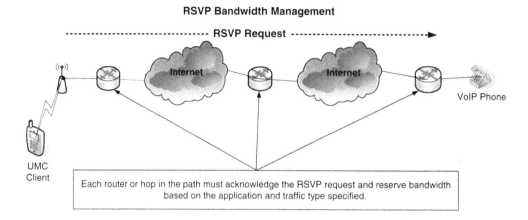

RSVP Bandwidth Management

RSVP Request

VoIP Phone

UMC
Client

Each router or hop in the path must acknowledge the RSVP request and reserve bandwidth
based on the application and traffic type specified.

Figure 8.3: An RSVP example.

The problem with RSVP is that its implementation is not universal across all possible
Internet routers or network subnets. For this reason, it is not likely that RSVP has a
major impact on a UMC application outside a corporate management domain.

8.3   Converged Media Prioritization

Management of congestion by bandwidth reservation has little impact outside the
corporate IT domain, but there is hope for those desiring an optimized mobile
experience across the Internet: *media prioritization*. There are standards and
mechanisms in place that can ensure a real-time application like UMC will have
minimum congestion problems in most any locale.

Figure 8.4 describes the three major segments of a network to which media access
management can be applied and where application-level prioritization is critical. Each
technology class has defined its own media prioritization schemes that can be leveraged
for end-to-end optimization in a large network, to provide the right QoS for UMC
applications.

8.3.1   WLAN WME/WMM

Media access optimization was achieved by proprietary means prior to ratification of
the 802.11e and Wireless Multimedia Extensions (WME)/WiFi Multimedia (WMM)
specifications (see Section 6.3.5). Some physical-level access advantage can be

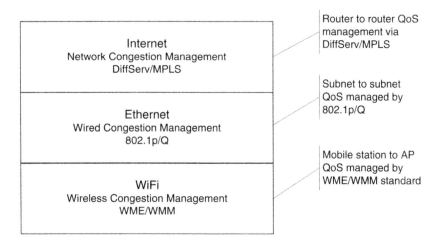

Figure 8.4: Layers of QoS services.

achieved by a real-time wireless application by modifying the carrier-sense/back-off algorithms. In such cases, the 802.11 standard specifies that a node desiring to transmit a frame but detecting a carrier must back off some 15 time slots before a retry. Using such an algorithm provides flexibility in gaining access to the media, even in heavily used networks, but some wireless voice products would implement an abbreviated back-off scheme to gain some slight advantage to transmit a frame. Products implementing such a scheme are typically voice-only devices and don't violate any of the 802.11 standard or WFA certification requirements. The most significant media access control, however, is realized when the handset and wireless infrastructure vendor supports 802.11e/WME and/or WMM.

Management of media access, without a control scheme, means that all wireless applications have equal chance of accessing the media for transmit/receive, regardless of the requirement of the specific application. In a lightly loaded wireless network, such schemes may not be necessary, and file transfers and VoIP applications can coexist. In corporate wireless environments, however, the network resources become the lifeblood of the business, and bandwidth is often a precious commodity that must be policed and managed based on the application-specific requirements.

A detailed discussion of these standards will not be provided here, but let's look at the essence of how these standards can be used to manage the media. The underlying principle of these standards is a method by which any wireless node can negotiate a guaranteed bandwidth from its current network access point. This is achieved through a dialog whereby the mobile device sends a request to the managing access point to guarantee a level of service. Designated as a TSPEC in the standard, this element can be assigned a scalar value indicating a requested bandwidth guarantee (seven levels, from *best-effort* to *voice/video* to *control* level). The receiving AP determines whether it can honor the level of service and responds with an acceptance or rejection of the request. A rejection might come from the fact that the AP has allocated all the associated noncongested bandwidth to other mobile units. In such cases, the requesting mobile unit has to make a choice of accepting a *best effort* on transmit/receive or roam to a different AP, where the congestion is less. Such conditions do exist in a live wireless network and complicate the sophistication of the layer 1/2 driver necessary to support these features (see Figure 8.5).

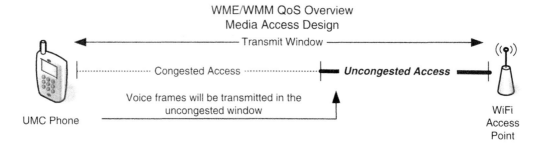

Figure 8.5: Congested/uncongested communications windows.

In some ways, a wireless terminal can be metaphorically described as a blind wanderer that must "feel" his way through a wireless forest. Once a mobile device "finds" an AP, the search for the next AP must go on, if the device is truly mobile. Today's mobile devices must periodically send out *probes* on different channels to receive responses to identify the presence of an AP.[2] Once discovered, the challenge becomes as to *when* to roam to the next AP, as triggered by weak signal strength, high transmit/receive errors, or lack of QoS. In general, roaming algorithms fall into two categories:

- *Desperation roams.* The launching of the roam operation does not occur until the current AP connection is no longer viable. Such a mechanism results in poor voice quality for UMC applications when roaming within a corporate WLAN.

- *Preemptive roams.* The launching of the roam operation occurs before severe deterioration of the current AP connection.

---

[2] Many enterprises disable sending "beacons" on wireless LANs due to security concerns.

For the optimal UMC application, the mobile unit should request a voice-grade TSPEC to obviate contention with traffic to and from the AP and preemptively roam to the best available AP when the link quality deteriorates. The speed of these roams may be optimized through conformance with the 802.11e and 802.11i standards.

8.3.2 IEEE 802.1p/Q

Support of prioritized traffic across a wireless domain is only the beginning. To propagate the desired prioritization onto the wired LAN, the wireless infrastructure must provide some method for translating from the equivalent QoS services on the wireless network to the wired network. This is accomplished by mapping the TSPEC level of QoS to an equivalent configuration on the Ethernet employing 802.1p. Coupled with the concept of VLANs is a concept of *user priority* that maps to the wireless TSPEC values. In an 802.1Q header (see Table 8.1) that is used to enable VLAN services, 802.1p specifies how a three-bit field is used to identify packet priority.

Table 8.1: 802.1Q header description

Header field	Field size	Description		
User Priority (IEEE 802.1p)	3 bit	Packet priority from voice, with top priority (0x111) to best effort (0x000)		
Canonical Format Indicator (CFI)	1 bit	Always set to zero for Ethernet but used for bridging Ethernet to Token Ring		
VLAN ID (VID)	12 bits	VLAN ID		

Frames so formatted are transmitted across the Ethernet and are given the appropriate priority at each router boundary (see Figure 8.6). The presumption, of course, is that all routers have been configured to provide this level of packet prioritization. Within a corporate LAN this is a manageable task, but across the worldwide Internet there is no guarantee that this level of prioritization is provided. Homogenous QoS will eventually be provided across the Internet, but this is still an evolutionary process.

The preceding are layer 2 options for implementing packet prioritization. Complementing these services are layer 3 prioritization schemas.

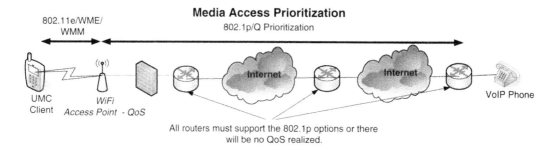

Figure 8.6: QoS via 802.11e and 802.1p/Q.

8.3.3 IP Type-of-Service (TOS) and Differentiated Services (DiffServ)

When the TCP/IP header was designed, there was an element (see Figure 8.7) defined to specify the *Type of Service* (TOS), which was intended to provide a method for *tagging* a frame for special-priority processing. It is the responsibility of the source node to establish the required priority, because it is coupled with the driving application.

The datagram fields are defined as:

- *Version.* The IP version number, currently v4.0. Version 6.0 will become a consideration in the next few years.

- *HLen.* Header length in 32-bit words.

- *Type of Service.* The way the datagram should be used, e.g., delay, precedence, reliability, minimum cost, throughput, etc.

- *Total Len.* The number of bytes included in the IP frame.

- *ID.* The frame identification is a unique number assigned to a frame fragment to aid in frame reassembly.

4	4	8	16	16	3	13	8	8	16	32	32		
Version	HLen	TOS byte	Total Len	ID	Flags	Frag Offset	TTL	Protocol	Header Checksum	SA	DA	IP Options	Data

Figure 8.7: IP datagram format.

- *Flags.* Bit 0 is reserved. Bit 1 can be fragmented (= 0) or cannot be fragmented (= 1). Bit 2 identifies the final frame fragment (= 1) or 0 if more frames follow.

- *Frag Offset.* Frame positional information for frame reconstruction.

- *TTL.* Time to live, or the number of router hops allowed.

- *Protocol.* Layer 4 protocol types: UDP, TCP, ICMP, IGRP or OSPF.

- *Header Checksum.* Header error control.

- *SA.* Source Address.

- *DA.* Destination Address.

- *IP Options.* Used for testing and debugging.

- *Data.* Balance of transmit frame.

Where the intent of the TOS bit was rather rudimentary, this 8-bit field has been redefined and used for Differentiated Services (DiffServ), which is a more comprehensive end-to-end media priority scheme across multiple networks. Network priority requests are embedded in each frame in what is now called the Differentiated Services Code Point (DSCP). This 6-bit value allows for up to 64 (i.e., 2^6) different traffic class types to be identified for special processing when crossing router boundaries. Such flexibility allows each individual frame to be tagged with regard to expected forwarding behaviors such as priority and even "assured" forwarding within any one class of frames. To be effective in managing end-to-end priorities, it is essential that all routers between the source and destination be configured in the same manner with regard to support of DiffServ. Unfortunately, within the Internet, there is no way to enforce such policies, and there is no absolute guarantee of consistent transport priority except within a single *DiffServ domain* that is under one management operation.

At each router boundary layer, different policies may be applied, even if DiffServ is enabled, when it comes to management of overall bandwidth allocation. Based on instantaneous traffic loads, a router may decide to ignore the DSCP based on upper load-level policies and forward any one frame on a *best-effort* basis. Therefore, it is impossible to predict the end-to-end priority forwarding behaviors when traffic crosses multiple router boundaries. Additionally, some development environments may not allow an application program the ability to establish any specific precedence or priorities for its traffic. For example, a mobile handset may be running a wireless VoIP application concurrently with a customer relationship management (CRM) database application. In the optimum case, all traffic from the VoIP application should have a DSCP suited for a voice application and the database traffic from the CRM application be tagged for best-effort. Unfortunately, no broad support for such granularity is provided to the application developer community. In many cases, the application has no control over the priority requirements for its own traffic.

In the long run, the easiest solution for the network congestion problem is to continually develop and deploy faster and higher-bandwidth network pipes that simply provide great capacity rather than managing priority over a restricted bandwidth limit.

8.3.4   Multiprotocol Label Switching

Multiprotocol label switching (MPLS) is an architecture used by service providers to manage bandwidth, network routes, and QoS in advance of traffic arrival for data associated with specific applications. In principle, this is a method by which layer 2 traffic priorities are managed at a layer 3 level across autonomous network domains. When a packet enters an MPLS-enabled network, each frame is analyzed and a *forwarding* wrapper (forwarding equivalence class, or FEC) is placed around it. As that packet traverses the MPLS-enabled network, each router makes a simple check in the new header and determines the next routing destination without having to perform any deep-packet inspection. By properly assigning appropriate FEC identifiers, MPLS tagged traffic can be management in a more straightforward manner to optimize for QoS and bandwidth assignments critical for the viability of applications such as VoIP.

Individual enterprise UMC customers will not need to deploy MPLS but should seek out service providers who can support this level of traffic optimization. The consumer UMC solutions will have some level of guaranteed QoS from the carrier and will not be a direct "buy" concern.

8.4   Network Congestion Management Readiness: A Summary

For consumer UMC users, there is little that can be done proactively to ensure optimal QoS through prioritization when they access the network (whether intranet or Internet). They are at the mercy of the management policies enforced on each subrouter segment.

For enterprise IT managers, congestion control on networks will always be a challenge. The old adage, "Infinite resources will be infinitely used," is true for today's networks. However, corporate IT managers can take actions to ensure that the best experience can be achieved on the networks they control. Implementation of wireless and wired QoS mechanisms across all network elements will go a long way toward guaranteeing the best UMC experience, if only on their managed network segments.

CHAPTER 9

Mobile Handset Solutions

9.1 Dual-Mode Handset Landscape

The technology lynchpin of any UMC solution is the commercial availability of a dual-mode (WiFi and cellular) enabled handset or terminal. These devices have been on the market since 2004 with such offerings as the HP iPAQ 6315 and others, but these early dual-mode devices were not positioned as UMC or FMC devices but rather as devices that were network *agile* on an application-by-application basis. The value of the WiFi services was viewed primarily as a higher-speed Internet access option rather than for any VoIP application. Such a perspective so permeated the design philosophy that audio directed through the WiFi connection was hard-routed to a back speakerphone and not to the front speaker expected for telephony use. Initial FMC developers quickly found that, with these devices, a cellular phone call audio would be played through the front speaker, as expected, but any wireless VoIP audio would play out of the back speaker. This behavior, of course, is not acceptable to a phone user.

Early dual-mode products were also weak in the support of WiFi connectivity. Often the embedded antenna design was not RF-sensitive enough, the effective connection range was significantly shortened, and RSSI values were inaccurate. Also, AP-to-AP roam logic was not optimized for voice. As pointed out in an earlier chapter, a voice application needs such roams to occur in fewer than 500 milliseconds or voice quality may suffer. WiFi driver architecture was often modeled after a standard Ethernet driver, and roam decisions were handled by *desperation*; that is, when the existing AP signal deteriorated to the point of a lost connection, the driver would then launch a "scan" to seek a nearby AP to roam to. This operation could take up to 30 seconds when scanning all 14 channels and waiting for potential AP responses on each channel. In many cases, VoIP implementations on such devices would drop the call when roaming between APs. It was quickly very clear that the functionality of the WiFi layer 2 driver required significantly more sophistication than was being delivered.

A third critical consideration for UMC-supported dual-mode devices was battery life. Power design and management have been optimized for cellular services. All users expect four to six hours of talk time and days of standby time on cellular phones. Using WiFi as a voice transport, however, was a different matter. Transmit power requirements for WiFi are significantly higher than for cellular radios; for this reason talk times were found to be sub-one hour and standby times measured in hours, not days. Such characteristics were a major hurdle for UMC adoption. What was missing was aggressive power management implementation in the drivers and chip-level enhancements of power design for the radio circuits. Fortunately, the second and third generations of dual-mode devices now sport such enhancements. Battery life is still a characteristic that can be improved, but the current commercial offerings are acceptable for most UMC users.

9.1.1   Major UMC Handset Manufacturers

Because UMC is a nascent market, initially there were few handset manufacturers that would boldly step into this market. Since wireless carriers were concerned about WiFi cannibalizing their *minutes* or *subscribers*, they were more than mildly reluctant to back dual-mode offerings. Indeed, in some markets, certain models of phones were introduced with dual-mode configurations while the same design was introduced into other markets with the WiFi radios absent. In some cases, when the WiFi radios were present, the network services were hobbled to only support half-duplex traffic and not the voice-required full-duplex. In this manner, the WiFi connection could be used to surf the Web but could not sustain a real-time voice connection.

Traditionally, the carriers have been the sole source for purchasing cellular phones; thus they were able to hold tight rein over the market. This market reality is, however, changing. End users may now procure *unlocked* (noncarrier-specific) dual-mode phones from noncarrier distribution sources. The European market was the first to become more "customer friendly" and to broadly support unlocked phones. Within the past 12 months there has been a shift in market availability for such devices in the North American market. Such shifts result in a more competitive environment, resulting in the end customer having lower-cost UMC device choices available.

There are still relatively few manufacturers that produce dual-mode handsets. The most aggressive in this market have been Nokia, Samsung, Motorola, Sony-Ericsson, and the Taiwanese High Tech Computer (HTC). Nokia has committed that going forward, all its cellular phones will be dual-mode and will have a full line of Eseries smart phones based on the Symbian OS that are fully 3GPP compliant.

Historically, HTC has been an OEM source of manufactured phones that were marketed under various brands, including UTStarcom, Hewlett-Packard, and Cingular. These were Microsoft Windows Mobile PDA-class devices and filled a niche in the market not taken by the Nokia offerings. HTC is now self-branding its phones and is expanding its dual-mode portfolio with consumer-centric smartphones such as the T-Mobile Shadow and T-Mobile Wing.

Other manufacturers of dual-mode devices are Motorola/Symbol, Intermec, Research in Motion (RIM), Sony-Ericsson, and others (see Figure 9.1). Each of these vendors has targeted some subsection of the UMC mobile communication market and sold through its channels.

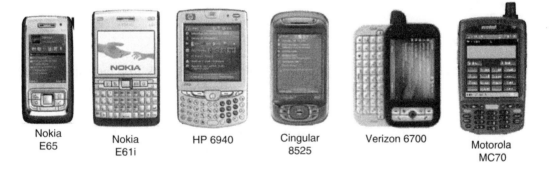

Nokia E65   Nokia E61i   HP 6940   Cingular 8525   Verizon 6700   Motorola MC70

Figure 9.1: Example of Commercially Available Dual-mode Phones.

Because these devices were designed as phones, they all have built-in dialers and standard cellular phone interfaces. Some even come preconfigured with a SIP-based softphone for use with VoIP applications. Few of them, however, come from the factory UMC ready. If a device does come configured to support UMC, it will most likely be a

3G-UMA/GAN device. Research in Motion has announced its first dual-mode device, the Blackberry 8820, being sold through a UMA/FMC-supporting wireless carrier.

One major challenge for vendors entering the UMC market is determining how to add the UMC functional components to the commercial phone. If the UMC vendor has not successfully partnered with the handset manufacturer to replace the native dialer, the challenge is to install an adjunct UMC dialer in such a way that it can coexist with the native dialer. Because the native dialer is tightly coupled with the OS, it is virtually impossible to replace, and UMC "add-ons" have to contend with working around certain functional limitations resulting from this situation. More details follow in the next section.

9.1.2   UMC Platform Challenges

Regardless of whether the dual-mode device is based on Symbian, Microsoft Windows Mobile, or Linux, they all seem to have generic functional challenges:

- WiFi: robustness and reliability

- Network flexibility

- Audio routing between front and back speakers

- Battery life

9.1.2.1   WiFi Robustness

How reliable and dependable the WiFi connection is relates back to a number of characteristics of both the hardware circuitry and the sophistication of the layer 2 MAC driver:

- *Antenna sensitivity.* The higher the antenna sensitivity of a device, the greater the effective coverage range that can be experienced. Most handheld mobile devices will never achieve the antenna sensitivity experienced with a laptop computer, but the effective sensitivity has a great deal to do with the maximized utilization of the WLAN. Erratic RSSI-level reporting has plagued early WiFi-VoIP implementations in managing voice quality. Handhelds that have an optimized antenna design will include a feature called *diversity.* This is a two-antenna design whereby transmit/receive interference is minimized by alternating use of the two antennas.

- *Roam agility.* Fast, secure roams between any two APs within a WLAN are important to ensure voice quality for the wireless call. The fast-roaming IEEE (802.11r) standard has been ratified but no commercial support is available as of the publishing of this book. Support for such features is embedded in the layer 2 driver logic.

- *Association robustness.* The ability to remain associated with a specific AP is critical, especially in an environment of high interference. This behavior is a characteristic of the MAC-level driver and varies from vendor to vendor. Lack of a solid, sustainable association results in too many roam attempts, which directly impacts voice quality.

- *Proper level of security support.* This characteristic has to do with conformance with both standards and proprietary-based wireless security solutions. For example, Cisco Systems has developed its Cisco Compatible Extensions (CCX) that supports a number of beneficial features that are Cisco unique and not specified by any standard. If a company has selected Cisco as the vendor of choice for wireless, CCX fast roaming and QoS services will be important as supported features on the mobile handset.

Each device will exhibit its own unique WiFi characteristics, and it is important that a company evaluate that specific device in the operating environment before a final buy decision is made. One of these factors may become a key decision metric in the final purchase.

9.1.2.2 Network Flexibility: Variances in Platform Design

Each operating system used for mobile handsets has its own unique approach for supporting, making, and breaking network connections, whether wireless or wireline. Most have a *communication manager* that is responsible for managing network connections, but there are design differences that may be important in a purchase decision from any one vendor. By intent, the communications manager is supposed to present an abstracted network interface that simplifies the operational management of network access. All require baseline configuration where all possible network connections are profiled (with proper security), including Bluetooth, WiFi, and USB connections.

Configuration management for WiFi will be the most challenging for UMC users. This is because there are:

- Corporate-sanctioned WiFi networks (corporate campus – behind the firewall)

- Remote private WiFi networks (home or remote office)

- Remote public WiFi networks (commercial WiFi services)

The first two classes of WLANs are sustainable because they are rather static regarding the users' nominal work week/day experience. These usually don't change with regard to ESSID or security and can be managed from a single corporate IT department. The third network, remote public WiFi, is the challenging one. To take advantage of these wireless services (in airports, hotels, malls, and so on), the individual end user must be able to configure these networks himself or herself. Most laptop wireless services present you with a list of accessible WiFi ESSIDs, and access is granted through use of a Web interface; the user enters license credentials or financial information for charges. Once a new wireless network has been configured, you're home free, right? Not so fast.

With most applications that are designed to operate in a wireless environment, once the network has been profiled and the device is associated, the application TCP services are made available to the application and execution is straightforward. With Symbian, however, there is an additional wrinkle. Whereas the majority of communication managers and applications have low coupling, in the Symbian "world" the application is tightly coupled to the specific network. Figure 9.2 shows an example of the differences

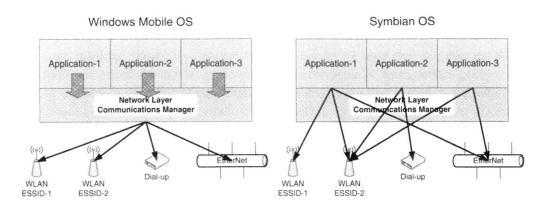

Figure 9.2: Communication manager design differences.

between Windows Mobile and Symbian in requirements in terms of how applications gain access to a specific network.

In the Windows Mobile world, once the network has been configured, its services are available for all network-based applications through a generic TCP/IP API interface. With applications running on a Symbian platform, there is an additional requirement that *each* application must have an *access point preference list* enabled to permit it to access any network service. This architecture has its downside in that an additional configuration step is required for every network application installed *and* if the applications do not have the exact same access point preference list, the mobile terminal will not have the same application environment support across all networks. This can be somewhat confusing for the user; that is, Application #1 works in network #1 but not in Network #2! Deploying a UMC solution on Symbian-based phones will require some additional network planning to optimize the usability of the mobile devices.

9.1.2.3 Audio Routing

Without exception, every first-generation dual-mode handset had the same problem: Audio traffic through the WiFi network connection was auto-routed to the back speaker. No configuration option was provided; with some designs, it was dictated by the electronic circuit in the handset. The presumptions of such designs were that (1) WiFi was only going to be used to play back audio from Web pages or (2) the audio would be routed to a headset. With such designs, a cellular call on a UMC handset would work fine with the audio being routed to the expected front speaker (as with any phone), but when that device roamed into WiFi coverage, the audio for the call would be routed to the back speaker. This is, of course, unacceptable for use in a wireless-VoIP scenario.

Handset manufacturers were informed of this shortcoming very quickly by all the UMC market contenders. Their response, however, was rather slow because the early upside market opportunity for UMC sales was too small. As the market has grown, handset manufacturers have acknowledged the need for application-directed audio-routing and have begun providing device-specific SDKs to be linked with UMC applications. The good news on this front is that the third and subsequent generation of dual-mode phones will not be plagued with this problem. The current dual-mode installed base, however, might not be blessed with such options. Many of these phones have been through end-of-life (EOL), and no future engineering resources will be applied.

In selecting a UMC solution, it will be important to know whether the application vendor has partnered with the handset vendor on this issue *and* has embedded the audio routing into its product.

9.1.2.4 Battery Life

Most consumers think little of the battery life on their cellular phone. The power demand for cellular phones has been greatly optimized over the past 20 years, and users are accustomed to purchasing automobile or laptop chargers for their phones. Cell phone batteries do go dead, but for most users, this is a manageable issue. With the advent of WiFi being added to the handset, power requirements for these devices have changed dramatically.

Transmit and receive power requirements for WiFi are greater than for typical cellular technology; this additional burden has strained the capacity of small, high-density batteries to the limit. With both cellular and WiFi radios enabled, a WiFi call is apt to completely drain a battery in less than one hour! Later generations of dual-mode phones do a little better on this metric, but they are nowhere near the battery life metrics for the cellular radio alone. To make a UMC solution viable over a long business workday, something has to change.

Optimization of battery life in a UMC device is achieved using two general solutions:

- *Implementing WMM power save.* This driver-level feature optimizes management of power demand on the mobile radio by placing the radio in a sleep mode when little or no traffic is required. Products that conform to this standard are certified by the WiFi Alliance as part of their WMM certification. This feature has a significantly positive impact on standby time for the mobile handset.

- *Higher-capacity batteries.* To ensure longer talk times on a UMC device, it is often the solution to configure it with a higher-capacity battery. Most dual-mode handset manufacturers offer these as accessories. They bump the cost of the total system a little and usually make the device a little heavier, but the increased usable time is the real benefit.

As the technology moves forward, the power requirements of WiFi components will become more efficient, thereby extending effective battery life. However, when evaluating a UMC solution, ask the vendor for battery life specifications for the device of interest. These should include cellular talk time, WiFi talk time, and several variants of standby time configurations.

9.1.2.5 Other Considerations

In selecting, purchasing, and deploying a UMC solution, other considerations that can affect the usability need to be addressed:

*   *Codec/audio encoding.* In most VoIP systems supporting a generic G.711 (64 Kbps), audio codec is not an issue. This is a reliable encoding standard that provides excellent voice quality over almost any network. There are other codecs (i.e., G.729, G.726, and ILBC) that provide different value benefits. Of most importance are the low bit rate (LBR) codecs that significantly lower the network bandwidth required for transporting an audio stream. Most popular of these codecs is G.729, which provides for excellent voice quality but requires only an 8 kbps bandwidth. Certain PBX and hosted providers no longer support G.711, and the winning solution must support a compatible codec.

*   *Professional and Personal modes.* Many UMC solutions require a mobile handset to support two phone numbers: one that the cellular carrier issues and one that is used by the UMC mobility authority (e.g., an office phone number). It is not unreasonable for a user to desire to make and receive business calls on the UMC number and personal calls on the carrier assigned number.

*   *Global positioning systems (GPS).* More sophisticated terminals and smartphones are coming on the market equipped with GPS. This technology is a double-edged sword in that it can aid the user in determining travel directions but can also be used by a business entity for monitoring the whereabouts of the individual. For example, Papa John's pizza delivery has launched the TrackMyPizza.com Website, where the customer can track the pizza delivery person. This feature in a UMC handset opens up many untapped opportunities.

*   *Push to Talk (PTT).* The ability to use a walkie-talkie mode on a cellular phone is *very* popular. This half-duplex communications mode allows a single person to send audio to multiple units simultaneously. Often used in construction, healthcare, and transportation and logistics fields, this feature has become a requirement for many who seek extended mobile communications. The challenge is implementing half-duplex over multiple networks. Some UMC vendors are promising PTT sometime in 2008.

9.1.3   Microsoft Windows Mobile UMC

By far, the most popular dual-mode PDAs are based on Windows Mobile (2005 or 2006). These have come from a number of vendors, but all have architectural roots that are limited somewhat in flexibility to support UMC implementations. The native dialer, *cprog.exe*, is implemented as a system service and therefore cannot be terminated or replaced, which poses some real challenges to UMC developers.

Without going into too many details, any third-party UMC application must be implemented to coexist with cprog.exe as best as possible. A UMC application can be installed that acquires the appropriate peripheral services (keyboard, microphone, speaker, etc.) but at a cost. As implemented, cprog.exe manages reporting the active cellular signal, and any attempt to suppress this thread from executing will result in an error in cellular signal strength reporting. There could be cases where a user has strong cellular signal but it is reported as low, or the reverse may be occur. Because cprog.exe cannot be "killed," it has zombie-like qualities in that, depending on the state of the handset, an inbound call may be intercepted by cprog.exe before the UMC application gets this interrupt. This scenario will cause two phone rings to occur: one from cprog.exe and one from the UMC application (see Figure 9.3)—confusing! Another artifact of the current design is that when the "dial" key is intercepted by the UMC application and it is terminated, that key binding is not returned to cprog.exe. This was a sequence that was never thought to happen. The only way to restore the default key binding to cprog.exe is to reboot the device!

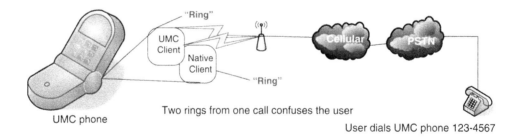

Figure 9.3: Resolving UMC/native dialer conflict.

These shortcomings may be addressed in future releases of Microsoft Windows Mobile-based products, but it is strongly recommended that the prospective buyer understand any vendor limitations with the product before making a final decision.

9.1.4   Symbian UMC

Nokia, along other vendors, invested in Symbian and chose it as the operating system for its cellular phones. Unlike Windows Mobile, Symbian is a compact operating environment tailored for smartphones and was not designed as a general third-party application-hosting platform. It does have provisions for installing third-party applications, but to guarantee that third-party applications are well behaved, such products must successfully pass a Symbian and Nokia certification test suite. Many Eseries Nokia phones come with individual factory-configured GSM and SIP dialer applications, but neither is UMC seamless roaming capable. Certain Nokia models (e.g., 6301 and 6136) are UMA ready from the factory. But non-3GPP-UMC applications must be added as an after-sale operation.

Like the Windows Mobile situation, any third-party add-on will have to play "nice" with the native dialer. Consistent behaviors have to be managed, such as being able to initiate a call from the *standby* display when a third-party UMC application is loaded. Without explicit third-party configuration, some calls may be made through the native dialer and others through the third-party dialer view; it can be very context sensitive. And as with Windows Mobile implementations, inbound calls to the device may be intercepted by the native dialer before the overlay dialer is invoked and the user may experience two different ringbacks.

9.1.5   Linux UMC

At the time of this writing there were not a large number of Linux-based dual-mode phones on the market. Companies such as G-Tek and e28 have been developing dual-mode Linux-based handsets, but they have not been launched in any broad markets. Regardless of the uniqueness of the operating system, most Linux dual-mode phones will have to be content with the same set of usability and network access problems as the other products.

Finding the right UMC solution (handset and OS) will involve some investigation of multiple-vendor solutions. At the time of this writing, the only seamless UMC solutions were those where the handset manufacturer partnered with a UMC vendor and had the backing of a wireless carrier. This model began market launch in 2007 in limited test regions because it also required the wireless carrier to upgrade its network in that area. However, each successive generation of dual-mode handsets and operating systems results in a platform more welcoming to hosting the broad spectrum of UMC-class applications from multiple vendors.

9.2   Determine Your Mobile Handset Requirements

Beyond a consumer candy-bar or flip-phone class UMC device, businesses are interested in the productivity advantage such technology will give their companies, but their physical requirements may be quite different. Requirements of multifunctionality, durability, and general form factor come into play when businesses are making UMC decisions. There is, fortunately, a broader selection of UMC-capable devices coming to market to aid the enterprise in making the best-in-class selection.

9.2.1   Rugged vs. Nonrugged

Small, candy-bar phones usually do not have a long life when deployed into an enterprise environment. Communication in warehouses, super-center stores, and outdoor venues requires phones that are more durable than standard cell phones. For many industrial uses, UMC device candidates will have to "muscle up" and be ready to take the wear-and-tear of these work environments. Fortunately, a few vendors are coming to market to meet these requirements; companies such as Motorola and Intermec have dual-mode devices on the market that can survive up to six-foot drops onto concrete; really durable! Such devices do come with a premium price and are often configured as multifunctional terminals rather than small pocket-sized phones.

9.2.2   PDAs vs. Smartphones

The two most popular form factors for UMC devices are personal digital assistants (PDAs) and smartphones. The primary difference between the two is that PDAs have a touch screen and QWERTY keyboard and smartphones do not support a touch screen and have a standard telephone keyboard. The PDA is friendlier for use with messaging applications and navigating through mobile corporate applications; smartphones are better for use as, surprise, *phones*. Selecting between the two form factors is not a cost issue as much as a purpose issue. If the mobile user is more bound to messaging and data-centric applications, a PDA UMC device would be appropriate; otherwise, a smartphone would be the choice. Most dual-mode handset manufacturers offer both form factors, making such decisions somewhat simpler and allowing you to stay with a single vendor regardless of form factor.

9.2.3   802.11 Support (a, b, g, n) Considerations

Deciding which 802.11 technology is needed may be a harder question to answer. Most WiFi-enabled devices on the market today are 802.11b/g devices supporting 2.4 GHz frequency ranges and upward to 54 Mbps data rate. For the vast majority of companies seeking wireless communications solutions, this level of technology is quite adequate and cost effective. The harder question is finding a WiFi solution supporting the 802.11a solutions. As of the first half of 2008, very few handsets and terminals on the market support this IEEE standard. Typically, these are a rugged multifunction terminal format and carry a high selling price.

The promise of very high data rate (>300 Mbps) that IEEE 802.11n brings will not be found in a handheld mobile terminal in the near future due to the higher cost and form factor of the supporting chipsets, the requirement for multiple antennas, and power demands for RF support. This technology must mature and be able to resolve the mobile handset-specific problems of cost and battery life. The future looks bright for a UMC 802.11n terminal but most likely not until 2009 or 2010.

CHAPTER 10

Hotspot and Hotzone Access

The versatility and convenience of wireless Internet access have expanded beyond the home and office with the commercial availability of low-cost WiFi products. Every home with an Internet connection can now support its own home hotspot for under $100 with high-speed (54Mbps) WiFi routers. As with any technology, the business world began to investigate how it could be used for profit, and commercial hotspots began popping up several years ago. Initial WiFi/Internet hotspots were deployed in coffee shops and airports, where the strategy was to attract customers to spend more time in the establishment (i.e., buy another cup of coffee) or where there was a captive audience that needed immediate communication services.

It didn't take long before large numbers of businesses saw financial opportunity in providing some kind of wireless commercial hotspot service. After all, today's citizen wants to be connected at all times, even when mobile. Wireless service providers have partnered with hotels, shopping malls, sandwich shops, convention centers, and other businesses to deploy hotspot service as a revenue-generating strategy, which analysts predict will grow at double-digit annual rates, to some 700,000 hotspots worldwide, by 2009.[1] Whether charged fees on a daily or hourly basis or whether free to patrons, subscribers or guests may access these services for email, text messaging, or Web surfing virtually anywhere. *Convenience* is the word, and people are willing to pay for it.

When planning for hotspot UMC use, there are some considerations that must be applied because of the access complexity that can be imposed at each different hotspot service (see Figure 10.1 and Table 10.1). There are limitless configuration

---

[1] Seventy-six percent have used WiFi at home, but less than 10% have used municipal hotspots, so the jury is still out (Forrester Research, Nortel report).

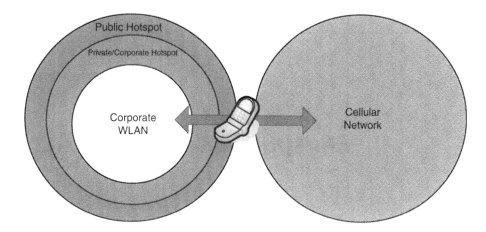

Figure 10.1: WiFi/hotspot network domains.

possibilities that define access policies for these hotspots, and it is important
that the UMC mobile client be flexible enough to accommodate support of
multiple hotspot classes. These range all the way from the completely free and
unrestricted home WiFi access to the subscription-based, user/password-
authenticated hotspots.

Even when hotspot accessibility technical hurdles have been cleared, the usefulness
of hotspots for UMC applications may be spotty. Most certainly, home and office
hotspots (a.k.a. WLANs) will be the most valuable because of the tight integration
to a home or office network infrastructure, and they are located where we spend
most of our time. Commercial hotspots at hotels, malls, convention centers, and
airports will be of more value to the road warrior because they're located where
the mobile worker tends to be. Municipal hotspots may be opportunistic because
of spotty assurance of bandwidth and dependence on the mobile user's destinations.

Leveraging municipal and public hotspots also places a configuration responsibility
on the end user that may be beyond his or her knowledge base of networking. A user
would have to recognize that a hotspot was deployed in the user's current location and
would have to define a *network profile* on his or her mobile device to use the resource.
For example, few UMC users will be able to take advantage of many metropolitan
hotspots because they would spend very little time out of doors, within range of such
services. The value of a hotspot to UMC users will depend greatly on locations where
they spend the most time.

Table 10.1: Hotspot classes

Hotspot class	Characterization	UMC value	Notes
Home/Residence or remote office	Sporadic authentication or encryption,[2] may be firewalled	High	Mostly used for family email and Internet access purposes
Commercial – fee based	Web-centric authentication for fee-based use, no encryption	Medium	Found in hotels, apartment complexes, coffee shops, airports that use WiFi to attract and retain customers for their primary business
Commercial – free access	No authentication or encryption	Medium	Found in hotels, apartment complexes, coffee shops, airports that use WiFi to attract and retain customers for their primary business
Commercial – free access w/commercial play	No authentication or encryption – only acceptance of access polices	Medium	Offered as a free service for specific usage periods in exchange for viewing product commercials
Municipal/public	Web-centric authentication, no encryption	Low	Provided as a service to local residents meant to catalyze city business to raise the tax base. Can also be used for public safety purposes
Business/ Enterprise	Strong authentication, strong security, firewalled	High	Not provided as a service for public access but an extension to the corporate hardwired network. Public access may be provided as a courtesy through an open VLAN segment without encryption or based on a one-time credential assignment.

[2] Novice home users may not enable security at all, or the latest generation of WiFi products may enforce a high level of security automatically.

10.1   Wireless Internet Service Provider (WISP) Access Considerations

Corporate campuswide wireless connectivity to databases or application servers has become commonplace. The mobile enterprise worker is no longer tied to a desk but can now be productive in those long management meetings by responding to email (we all do it) while fulfilling corporate commitments in the lunch room, conference room, and engineering lab. Deployment of WiFi within the enterprise has extended the reach of the network to virtually every corner in the corporate facility and has had a positive impact on the overall productivity of people in the enterprise.

Telecommuting is also a trend that has leveraged wireless freedom in the home or remote office. Just as wireless mobility has been provided to on-campus associates, remote workers can enjoy wireless freedom in their personal work locale. Though deployment is not as straightforward as on-campus wireless, remote wireless access can be installed with the cooperation of the corporate IT team and typically requires imposing supplementary security measures in the form of a VPN. The remote user's experience of wireless freedom can be almost identical to that of the on-campus associate.

As WiFi is embraced by our society, the phenomenon of public hotspots has emerged and is increasingly prevalent in urban areas. Major cities such as Philadelphia, San Jose, New Orleans, Mountain View, Santa Clara, San Francisco, and others are in the process of investigating public hotspot services under the sponsorship of the municipal authority. For the highly mobile worker (executive, field service, transportation, public safety, and others), these trends open up new potential for extending UMC connectivity. With the availability of this expanded wireless network service, what is the upside for leveraging these network services, and what are the hurdles to overcome so that we are able to use them?

10.2   The Hotspot LANscape

Hotspots were initially deployed to attract consumers to a commercial establishment with hope of increasing revenues through customer "stickiness." In a coffee shop, it was thought that a customer searching the Internet or working on email would drink more coffee and perhaps purchase a meal as long as the wireless connection was reliable and cost was not a major issue. These assumptions have proven correct: Wireless services *do* attract and retain a certain class of customer, and hotspots have popped up all over—tens of thousands of sites worldwide. Some hotspots are free, others are subscription based.

10.3   Hotspot Use Models

In the initial hotspot user model, an individual customer would pay a fee (per session, by day, or by subscription) for accessing the Internet at a business-hosted hotspot. After associating with the local WiFi Access Point, the user was required to launch a browser; the WISP hotspot provider would highjack this request and display a "login" Web page, referred to as a *walled garden*, where the user would enter the appropriate authentication information. Once authenticated, the user would receive the originally requested URL and continue surfing uninterrupted. Seems simple enough for the personal user, but is this sufficient for the mobile enterprise worker?

Hotspots (or *hotzones*) introduce additional factors the mobile worker must consider. Where a home or office WiFi service is "free" through conformance to only the WiFi security policy, access through public hotspots poses additional hurdles in gaining Internet access. Unlike most home or office WLANs, hotspots are always "open," which means that the WLAN signature (Extended Service Set ID, or ESSID) is broadcast with no imposed wireless encryption. However, a public hotspot is typically operated by a third party called a Wireless Internet Service Provider (WISP), and some kind of authentication process is typically imposed. Access to such wireless services is typically based on some acceptance of licensing domain or user identification and password (see Table 10.2).

Table 10.2: Hotspot access schema options

WISP access protocol	Authentication/ authorization	Cost	Description
Open access	None	Free	Some commercial sites such as hotels offer free WiFi service
Open access— commercial play	None	Free	Requires acceptance to view product commercials as part of the access policy
Sanctioned access	Accept use license	Free	Some hotspots require acknowledgment of usage agreement
Subscription access	User ID	Subscription or free	User enters a registered ID string (name or email address)
Secure subscription access	User ID and Password	Subscription	User enters a registered name and password

Hotspot support for a UMC application poses some unique challenges. However, if I'm on a cellular phone call with a UMC device and I walk into a WiFi hotspot coverage area, it defeats the purpose of "unbounded" mobility if I have to first launch a browser before I can hand over the call to the WiFi service; this process must be automatic. Early FMC offerings avoided this problem, supporting only home services, which have no such service access requirements.

Most consumer hotspot users browse the Web and answer email with little to no understanding of the business implications regarding security policies or application type. Besides requiring WISP auto login, enterprise users have more stringent security needs because they might access corporate data and use real-time applications such as VoIP. These requirements place a different set of demands on how and when hotspot services are to be accessed.

As the UMC market matures, the trend will be for WISPs and handset manufacturers to acknowledge the requirement for automating hotspot access and include this functionality to seamlessly log in to public hotspots. Several companies, including T-Mobile, Skype, and Nokia, have already stepped out with hotspot support. Also, consumer-focused companies like Devicescape and Boingo sell seamless hotspot login functionality for Nokia Eseries and Nseries dual-mode phones that is offered on a subscription-based pricing model. Boingo Mobile is marketed to the consumer segment as a program with a flat-fee structure providing dynamic access to the thousands of national and international hotspots where Boingo is the provider or has a roam agreement. Devicescape is more of a hotspot *access broker* in that a user must define her hotspot environment, free and subscription-based hotspots she wants to access, and will perform a proxy login for the user for any defined hotspot. Other major WISPs, such as iPASS, are investigating their play in the UMC market as an extension to their legacy enterprise hotspot market focus. To achieve maximum mobility, any serious UMC user will require such a service supported on his handset.

10.4   Impact of Convergence

The final hurdle for enterprise access through hotspots is application specific. Most enterprise applications are not real-time applications that require sub-300 millisecond responses, but this all changes with a UMC solution. Seamless network transitioning means that a phone call that began on the cellular network could be transitioned dynamically onto the Internet when the device came in range of a WiFi/Internet service.

For the roaming transition between cellular and WiFi networks to succeed, authentication, IP addressing, and security must happen automatically in the background, *with no user intervention*. This places additional requirements on evolving applications to manage security access, application access (if required), and WISP access in a transparent manner. Automatic authentication is an evolutionary enhanced service for this market that will become more readily available as demand for wireless real-time functionality (voice and video) becomes more widespread. Billing plans will also evolve to accommodate bandwidth demand-based schemes versus any elapsed login time design.

10.5   Hotspot Security Considerations

Placing a phone call from a public hotspot WiFi raises the question of security. Without an imposed WiFi security policy, any traffic between the handset and the hotspot access point has the potential to be monitored, recorded, and replayed. Most consumers don't mind if that happens, because their cellular calls are not necessarily secure from eavesdropping. Depending on the specific deployment of UMC, up to four levels of security can be imposed when you're attempting remote access (see Figure 10.2).

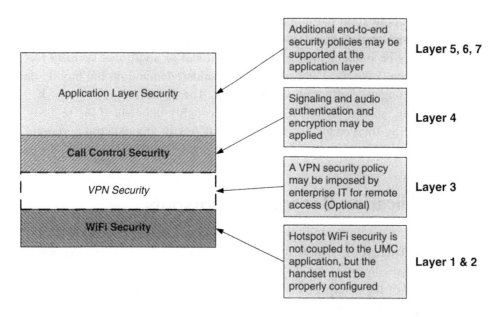

Figure 10.2: UMC security architecture.

Most certainly, there may be authentication and data encryption applied to the WiFi handset-to-access point link, but public hotspots are "open" and enforce no security. At a higher OSI layer, many corporations impose a virtual private network (VPN) policy to ensure an authenticated and encrypted end-to-end connection between the mobile unit and the hosting network. Imposing such a security measure on mobile handsets poses several operational problems:

- *General availability of a mobile VPN product.* There have been few VPN products on the market that will provide support for multiple OS platforms (Windows Mobile, Symbian, or Linux), which can provide the broadest mobile device options.

- *Sufficient CPU bandwidth to support good voice quality.* Many commercially available mobile handsets don't have sufficient CPU clock speeds to support the overhead of VPN processing without impacting the resulting voice quality. The added latency debt can have a severe impact on the acceptability of the voice application.

There are additional security options that can be implemented in place of a VPN that provide application-specific end-to-end security. The call control (SIP stack) can self-impose an Authentication-Authorization-Accounting (AAA) function that will fully authenticate the application participants and impose some kind of encryption on signaling and audio traffic. Standards like Secure Sockets Layer (SSL) and Secure RTP (SRTP) may be implemented to provide optional or additional security for a UMC application. Like the VPN solution, any added demand on the mobile device CPU can have a negative effect on voice quality. The process to select a UMC solution, therefore, should include a study to thoroughly understand the security options provided by the vendor and its conformance with any existing corporate security policies.

10.6   Application Barriers for Hotspots

Figure 10.3 depicts the three major network topologies that can be used to deploy wireless access. Though the *remote office access* configuration is more complex than the straightforward *enterprise access* schema, both share a common feature: They are under the control of the enterprise IT. End to end, a hosting enterprise has control over the security, bandwidth, and applications used within the confines of the enterprise network.

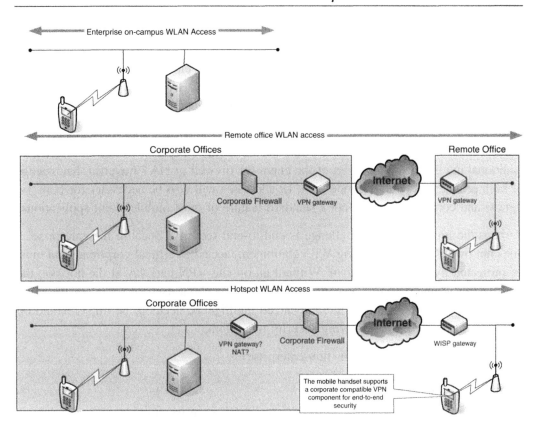

Figure 10.3: Wireless access models.

The third topology, *hotspot access*, raises additional complexities and hurdles that need to be addressed to provide the same seamless experience as the other two topologies. Because the hotspot is not under direct user or corporate control, these are considered a no man's land with regard to conformance to access, authentication, and security. To protect corporate information (voice and data), the application, under control of the business, must be able to apply sufficient access and security policies outside of any provided by the hotspot.

One of the first challenges is WISP access. Some Fortune 1000 companies have contracted with hotspot vendors or aggregators[3] to provide services to associates when they are outside the office through a corporate subscription service. Integration of a WISP

---

[3] Vendors that broker services for regional and national hotspot service providers.

service is not a major technical hurdle, but there are management and logistics problems that need to be solved. To utilize the hotspot services, the client must provide the proper authentication information (i.e., user name and password), either automatically or through user prompts. Manual entry of this information through a user interface is not a problem unless there is a requirement for real-time application support (discussed shortly).

Once access to the network has been granted, the user is assigned an IP address. This process too is straightforward, but communication with a corporate network may require additional intelligence on the part of the client for firewall or NAT traversal. Knowledge of both private and public service point IP addresses will also be necessary for both remote and corporate access—not a standard feature of most mobile client applications.

Perhaps the most important challenge is end-to-end security. Because the enterprise does not own the remote hosting WiFi equipment, access to critical corporate data must be protected by some mechanism. Without an on-site VPN gateway at the hotspot, the responsibility for supporting a secure link falls on the client itself or some collaborating WISP service module. In this case, advanced intelligence needs to be implemented on the client to perform the following security operations:

1. Detect a "foreign" WiFi (non-enterprise).

2. Execute a WISP service login.

3. Invoke enterprise-sanctioned VPN or IPSEC security policies.

With such an application model, corporate data is as safe as within a remote office. The distinction in this application is that without a WISP security function, the client itself will be responsible for security management.

10.7   Future Wireless Freedom in Hotspots

Projections for the hotspot market are rosy, with some analysts predicting that the number of public hotspots will exceed 200,000 by 2008. Additionally, over half of all mobile professionals have tried today's hotspots—in hotels, airports, and malls. Going forward, free hotspots may be supplanted by "free with advertising" (like TV) hotspots or advertising-free hotspots based on subscription. Providers with the largest geographic coverage will win the race for the enterprise hotspot business; much like pioneers in the Oklahoma land rush, entrepreneurial WISPs will rush to deploy WiFi coverage in metropolitan and commercial areas first and then target WiFi for the enterprise itself.

10.7.1   Municipal Hotspots

Once the initial 802.11 wireless security questions had been addressed, there was a rush to evaluate the value of providing municipal WiFi services. Philadelphia was the early leader in planning to deploy WiFi in this highly dense urban area. The philosophic perspective was that government-sponsored wireless Internet access would catalyze a general resurgence of downtown businesses, would raise the value of the tax base, and could also be used for public service purposes. As of April 2007, Philadelphia had 135 square miles of 802.11b/g WiFi covering the core of the city. Following suit, a number of other North American cities began serious investigations into implementing a municipal WiFi service, and some cities have launched such pilot services, but it has become apparent that the costs of deployment and maintenance are not offset by any increase in the urban tax base; the ROI is not realized. Though the jury has not rendered a final verdict on municipal WiFi, it is unlikely that it will be aggressively deployed, and there are major questions as to whether or not such open networks will provide sufficient bandwidth to support resource-intensive applications such as VoIP.

There are some technical questions facing municipal hotspots that may muddy the waters concerning any widespread deployments of which RF technology to deploy. Many municipal WiFi services are based on WiFi-Mesh technologies (802.11s), which will impose additional latency on real-time applications (voice/video) because of the wireless-to-wireless routing nature of mesh. It works well for applications that are not latency dependent, but there is no standards-based approach for prioritizing voice or video traffic traversing a WiFi mesh network. Also, looming on the horizon is WiMAX. This technology is perhaps better suited to *last-mile* hotspot services, but the problem with this technology is that the *mobile* WiMAX standard has not been ratified and there are questions as to when WiMAX handsets might be commercially available. Given that WiFi will continue to dominate indoor wireless, WiMAX may be a strong presence for last-mile services, and cellular networks remain, this means that there may be the potential for a tri-mode radio handset somewhere in the future.

10.7.2   Federated Hotspots

One interesting wireless market phenomenon was the formation of FON.[4] FON is a social federation of individuals who share their home or small business WiFi/Internet service. It has funding from major market players such as Google and Skype/eBay and

---

[4] www.fon.com.

has a worldwide organizational presence. Individuals can sign up as Foneros and share their home/office WiFi with other Foneros; the intent is to allow Foneros to travel globally and have free access to WiFi supported by local Foneros. Tools have been created to simplify locating regional FON hotspots through standard (and popular) Web interfaces. There are even provisions whereby a Fonero may generate some small revenue for allowing a non-Fonero access to his or her wireless LAN. The FON federation operates a billing arbitrage service and can bill these customers and revenue share with the participating Foneros. Load balancing and security management of a FON hotspot are achieved by deploying the FON wireless router: La Fonera.

The concept of FON is a good one, but the general availability of collective services remains geographically sparse. The variability of bandwidth and observed QoS may hamper any successful leveraging of the FON infrastructure for serious UMC purposes. The concept of wireless federations, however, is valid and may result in an eventual spread of WiFi access services.

10.7.3   Portable Hotspots

One approach to ensuring that you have WiFi access when you're out of the office is to carry the office WiFi with you. One WLAN vendor has announced a "remote access point" that can travel with you. When an Internet connection is available, this device may be connected and provides a secure VPN connection back to the office network that is wirelessly accessible from the hotel or meeting room. Of course, problems may exist with potential RF interference in public areas and inconvenience of the added weight to carry the portable AP, but such devices will have their applicability in certain market segments.

10.7.4   Hotspots: A UMC Viability Summary

A robust UMC application will require a reliable, secure hotspot with ample bandwidth, to be of any practical use. Management of UMC-friendly hotspots must, in part, follow the same configuration rigor demanded of an in-house wireless LAN. Erratic bandwidth availability and weak RF coverage will eliminate any one hotspot from properly servicing a UMC application. What this means to someone considering deploying a UMC solution and wanting to take advantage of public hotspots is that some investigation and planning must be performed to identify the best hotspot candidates that can best guarantee the level of service desired. The most reliable of the candidates will be those sponsored and managed by some WISP rather than a private or municipal service.

10.7.5   The Value Proposition for Hotspots

A major success factor for hotspots, beyond just the availability of the wireless handsets and RF infrastructure components, will be the pricing models. Consumer and corporate buyers will have to find the offered pricing models appealing to make hotspots financially successful. Like any market opportunity, there are numerous options for creative pricing models based on the type of target customer and the need of the service provided. Seriously mobile consumers and corporate road warriors will be attracted to an annual or month-by-month offering. For the sporadic nomad, a single-use plan, modeled by many hotel hotspot services on a per-day charge, would work. The diversity of hotspot pricing models will complicate corporate plans for hotspot budgeting, but judicious investigation of the corporate requirement for connectivity and the selection of the most geographically available hotspot services will significantly extend the connectivity of a UMC user.

10.8   A Bright Hotspot Future

Like the public telephone of the past, public WiFi hotspots seem to be here to stay. Their presence, however, must be justified by value service that is provided to a buying public voting with the dollar. The individual consumer will achieve some additional conveniences when leveraging a hotspot service, but enterprises stand to reap the greatest benefit. Mobilizing the enterprise beyond the campus meets the need of today's mobile worker for accessibility regardless of locale. Seamless roaming between cellular and WiFi networks with a host of enterprise applications that include VoIP and video is the Holy Grail. In effect, this can mobilize a worker's application environment, maximizing personal productivity. Availability of UMC solutions will allow the mobile worker to take productive advantage of this newfound unbounded freedom.

CHAPTER 11

Security Considerations

11.1 UMC Security Considerations: A View of the Landscape

Most consumers are naive about the security of their wireless phone calls. In the early days of analog cellular service, some knowledgeable engineer with a frequency-tunable RF receiver and some method to record the traffic could record and play back cell phone calls at will. That's most certainly a scary thought for most 21st-century security-conscience individuals. Because UMC potentially traverses multiple network topologies, different classes of security must be considered. It is important to understand the breadth and strength of any security services that are provided by the UMC solutions, and for enterprises it is important to understand how those security features might comply with corporate policies.

The level of security required by an individual consumer will most likely be much less than that required by a UMC enterprise solution, but all must have some level of security. The security motivations of a consumer and an enterprise solution will be different. Eavesdropping on a personal call would typically have no associated monetary impact but would still be an invasion of privacy. Eavesdropping on a business call has greater legal and financial ramifications. For example, information shared on a business call between a stock broker and her client would contain sensitive trading information that may be covered under the Sarbanes-Oxley Act, with defined legal consequences. The same is true of a conversation between a doctor and his patient, which is covered under the Health Insurance Portability and Accountability Act (HIPAA) statute.

A security service usually involves two major components:

- *Identification of the communicating parties.* Either two-way or one-way identification of the parties is required. Providing information to an unknown entity is a definite security breach. Identification of a user can be made by user name or password, biometrics (fingerprint, retinal scan, or voice pattern), or some other scheme. Once the user is identified (authenticated), the user link is established (authorized).

- *Encryption of the exchanged information.* There are multiple standards methods that can be applied to encrypt data. Some may impact an application such as voice by adding to the end-to-end latency, but the security risk will dictate the sophistication of the security required.

With a UMC product, there will be options to apply multiple, complementary security systems; understanding these options will be important in selecting the most secure products. This chapter specifically looks at the security options that are available and can be applied to the various technology components that make up a UMC solution.

11.2 Cellular Security: An Overview

In today's cellular world, security is embedded within the system, and the user is not required to take any action when attaching to the mobile network. Authentication is performed not on the basis of the user but rather on the device. Once past the device authentication phase, a security key exchange protocol sequence is completed, and all signaling and audio traffic from the mobile handset through the mobile network is automatically encrypted. In such a state, this phone may be used by any person, but all information transmitted or received will be secure.

11.3 WLAN Security: An Introduction

Utilizing the installed 802.11 WLAN for both data and voice applications makes a lot of sense. For consumers and businesses alike, availability of WLAN technology seems a perfect fit for our ever more mobile lifestyle.

For businesses, mobile wireless UMC connections to the PBX offer some real productivity enhancements not possible with legacy "deskbound" telephony products. However, if you've read any of the early articles about wireless LANs, you are acutely aware of well-documented and widespread security concerns. The initial RC4-based encryption scheme adopted as part of the IEEE 802.11 specification was Wired Equivalent Privacy (WEP). When WEP is enabled, all packets transmitted between a WLAN access point and mobile units are encrypted with a fixed 64- or 128-bit key.[1] In the early stages of the WLAN deployment, most users felt secure in knowing that the information being transmitted on their WLAN was encrypted and secure. So unconcerned was the WLAN community about wireless security that many WLAN sites had no security enabled at all. It was the seminal RC4 analysis[2] article that signaled the weakness in the WEP architecture and the fact that hackers, with the right tools and enough patience, could crack even the highest level of WEP security encryption.

WLANs open up a new world of untethered capabilities and enhanced mobility, but the threat of valuable data being stolen right out of the "air" made most enterprise CIOs pause when considering deploying a production WLAN. Industry backlash at the apparent weakness in the WEP architecture forced WLAN vendors and the IEEE 802.11 working group to go back into session and redefine new, more rigorous wireless security schemes that would be embraced by the market.

The IEEE standards body, industry alliances, and individual WLAN vendors responded by proposing multiple options for more rigorous and "bulletproof" security schemes. Prior to the ratification of the WLAN security standard (802.11i), most WLAN vendors offered "safe" WLAN solutions with some form of enhanced proprietary security that attempts to address WEP's weaknesses. The availability of support for 802.11i allows wireless LAN customers to breathe a sigh of relief with strengthened security, but there may be repercussions with regard to support for specific WLAN applications. The following subsections explore the challenges posed in implementing a secure UMC solution.

11.3.1 Security Concerns: Customer Responses

Facing the documented security problems with WEP, enterprises that had already deployed an 802.11 wireless LAN took many different steps to maximize security within the established standards framework. This included imposing alternate security policies

---

[1] Strictly speaking, a 40-bit or 104-bit sequence is used for the encryption with a 24-bit value for packet sequence numbering.
[2] "Weaknesses in the Key Scheduling Algorithm of RC4," Fluhrer, S.; Mantin, I.; and Shamir, A., *Eighth Annual Workshop on Selected Areas in Cryptography* (August 2001).

that minimized the overall security risk of running a production WLAN. Besides imposing the use of 128-bit WEP, other corporate policies have been enforced, such as:

- Virtual private networks (VPNs)

- Virtual LANs (VLANs)

- Access control lists (ACLs)

- Disabling broadcast ESSID

Even though the WEP/RC4 security standard is flawed, it still takes some minimal effort of tracing traffic to crack a static WEP key. iLabs[3] reported that it could take a relatively long period of "sniffing" on a heavily loaded network to get enough of the weak initialization vectors to crack a 128-bit WEP key. However, in any secure deployment of a wireless LAN, WEP should not be used.

Deploying an off-the-shelf VPN does enhance the security of data because it is an end-to-end encryption scheme, which impacts more than just the RF security domain. A large number of corporations have taken this approach because it is commercially available, is already under their IT department control, and provides security to satisfy their concerns about wireless hacker intrusion.

Isolation of voice traffic using VLANs is another approach to eliminate or minimize any security policy impact on a voice application. If a corporation deems that voice traffic is at low risk for sabotage, it can limit all voice traffic on the network to specific, nonsecure devices by placing all "voice" traffic on its own VLAN segment within the network. Additionally, most current WiFi access points support the concept of VLANs, which further simplifies use of the wireless infrastructure by having different security policies for the various VLANs and segregating voice and data traffic. Typically, well-implemented VLANs have no impact on a wireless VoIP application due to the extremely low latency in VLAN switching.

Implementing an ACL involves configuring each access point within the wireless network with a list of authorized MAC addresses. This provides an inclusive list of all preauthorized devices that can associate with the network and blocks rogue devices from accessing the network. Though it potentially prevents rogue devices from access to the network, it also places a large management burden on the network administrator

---

[3] InteropNet Labs (iLabs), "What's Wrong with WEP?" *Wireless LAN Security Interoperability Lab Report*, Series #5.

to keep the access points up to date each time a device is added or removed from the pool. Additionally, it is fairly simple to reprogram the MAC address of a system, and you can spoof a valid MAC address. It also doesn't prevent illegal use of a valid device on the network. There is, however, no impact on the effectiveness of a wireless VoIP application with respect to implementing ACL policies.

Another "security" policy often recommended by consultants is to disable broadcast ESSID. The purpose of this policy is to prevent the ESSID string from being placed in the beacon that was sent out from the access point. The theory is that no one using a wireless "sniffer" tool could detect the ESSID (sent in clear text) if it was not in the beacon frame, thus keeping secret the identity of the wireless LAN. The flaw with this approach is that, in turning off the ESSID in the beacon, you haven't constrained advertising the ESSID, because each client places the ESSID in its "association" request (or probe). Now rather than only the access point publishing the ESSID value, *all* the mobile clients publish it. So, instead of only having one device publishing the ESSID, tens or perhaps hundreds of devices would be broadcasting the ESSID, making it far easier to find the wireless network identity and thus *defeating the purpose of the policy*.

Regardless of the security policy enabled for the WLAN (WEP or 802.11i), further access control can be realized with the possibility of using VPN, VLANs, or ACLs or placing the wireless LAN on an unsecured portion of the corporate network and running only noncritical applications. The way these security policies and practices impact wireless VoIP applications will be discussed later in the chapter.

11.3.2 Security Concerns: Deployment Options

A VPN seems to be a straightforward solution to wireless security problems because it provides security control at the highest level: end-to-end encryption across the network. VPNs meet the requirement of authentication and data security but can impose a severe penalty on a real-time application such as VoIP. Without an assist from a high-end processor or coprocessor, applying the encryption policy at the transport level can degrade the resultant voice quality of a dual-mode mobile device. The "tunneling" of the packet flow through a VPN can also add to the overall latency of the system and further degrade voice quality.

It almost seems that VPNs and wireless VoIP are mutually exclusive. This is not quite true, but great care must be taken in implementing and deploying a wireless VPN solution. The computing power of the handheld battery capacity and the VPN design within the network fabric must all be considered in an attempt to guarantee a secure, high-quality VoIP solution.

Most of the industry is familiar with Cisco Systems' LEAP/RADIUS wireless security solution. In this architecture, each time a device reinserts itself into the network (i.e., roams), it must have full reauthentication via the RADIUS server. Depending on the complexity of the hosting network, such an operation can add 150–250 msec (or longer) of latency on a roam. Such a small fraction of a second is not significant to a data application, but it will result in a degradation of the voice quality with dropped packets when roaming.

11.3.3   Security Concerns: WLAN Best Practices

The wireless industry recognized the critical need for providing strong security services and ratified the 802.11i standard that defines a robust security standard for WiFi products. Out of this work came some new WLAN security options such as:

- WiFi Protected Access (WPA): Temporal Key Integrity Protocol (TKIP)

- Advanced Encryption Standard (AES)

These works are mostly derived from the IEEE 802.11i task group and are focused on defining two categories of wireless security schemes:

- Enhancing the standards-defined security that is compatible with current hardware products, that is, a security model that can be implemented without changing the hardware

- Defining a more rugged security standard that may require additional hardware to be built into future devices

Figure 11.1 identifies these new components and how they relate to the original security schemes. Implementations of TKIP, WPA, and WPA2 architectures have been defined to fit into category #1 of the new security offerings. AES is a more robust security algorithm that has already been adopted by the military and the federal government for their encryption standard[4] and is the driving technology for security category #2.

WPA is a security offering backed by the WiFi Alliance and is this body's definition of the way 802.11i's TKIP components can be implemented while assuring vendor

---

[4] AES has replaced the use of DES and Triple-DES in many government deployments.

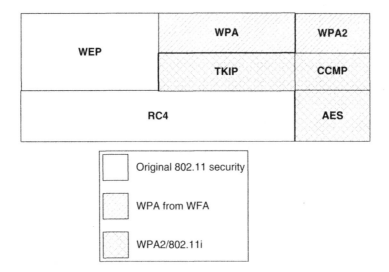

Figure 11.1: 802.11 security elements.

interoperability. All the TKIP elements of 802.11i (encryption, authentication, and message validation) have been included in the definition of WPA and guarantee interoperable wireless security schemes through firmware updates to older, commercially available, industry-standard hardware.

TKIP is an extension of the WEP standard that "plugs the hole" in the original RC4-based encryption standard. In this scheme, the scope of the key management scheme (Initialization Vector) has been significantly extended, along with a new requirement that each packet transmitted be encrypted with a new key. TKIP also includes the implementation of a *message integrity code* (MIC) that adds a per-packet source-validation mechanism.

The complete 802.11i (or WPA2 from WFA) standard defines Counter Mode with Cipher Block Chaining Message Authentication Code Protocol (CCMP) and uses Advanced Encryption Standard (AES) encryption, which is perhaps the ultimate strong security scheme. Generally accepted as the strongest encryption method available today, AES does provide stronger encryption services than RC4 but may also require hardware assistance when implemented on low-end, battery-powered devices. Implementing this level of the 802.11i standard may require new hardware platforms to be developed in order to provide optimal wireless voice quality.

11.3.4   Security End to End

Consideration of the security requirements for a mixed data/voice wireless network is critical to a successful UMC deployment because voice quality can be jeopardized by imposing overly stringent security policies. Therefore, it is important that businesses assess their risks with regard to wireless telephony applications and apply the minimum security policy to safeguard against that risk.

The advantages of a UMC solution can be lowered total cost of ownership (TCO), enhanced productivity of associates, and, possibly, lowered headcount due to a streamlined operation. The decision that must be faced is: "Do the advantages outweigh the risks?" Commercially available security options may vary from vendor to vendor, further complicating any final decision regarding wireless infrastructure and mobile units. Even with the implementation of 802.11i/WPA2, however, the final deployment decision will be made by defining configurations that balance among maximized voice quality, minimized security risk, and acceptable network management overhead.

11.4   Security Implementation Options

11.4.1   Call Signaling Security

For a UMC solution that employs SIP for setup and termination of calls, security of the signaling and audio traffic will be an important consideration. SIP sessions that are contained on the corporate internal hardwired LAN are not a major security threat because access to the media is controlled by physical access to the facility. Wireless SIP sessions, even on the corporate LAN, pose a different security problem. Implementation of WPA or WPA2 for security on the wireless link is essential, but many enterprises feel that even this is insufficient and place such components in the corporate DMZ. Decisions must be made on where to logically locate the iPBX interconnect. Placing this central component inside the DMZ can pose operational problems, including operating an internal firewall or NAT service. The presumption is that wireless will always be vulnerable to cracking or attack and that the hardwired LAN is not vulnerable.

SIP sessions that traverse the Internet pose a more difficult security challenge. Signaling and audio traffic originating outside the corporate firewall and traversing the Internet implicitly have a security exposure (see Figure 11.2).The wireless traffic from the

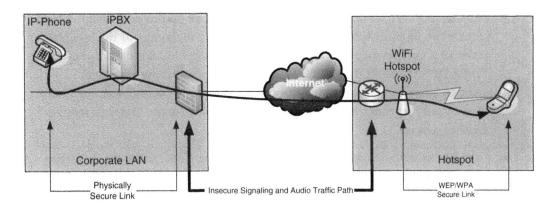

Figure 11.2: Remote UMC hotspot access without security.

mobile handset to the hotspot access point may be secured (although probably not in a public hotspot); the unknown path through the Internet exposes this application to all kinds of security problems.

The path of the VoIP traffic crosses unknown numbers of routers and through service provider media gateways and session border controllers. At any point in this journey, the traffic may be recorded or even hijacked for nefarious purposes if it is not encrypted. Only an end-to-end security design will allow for the maximum mobility with the fewest security risks (see Figure 11.3). To achieve that end, a few usage models are available:

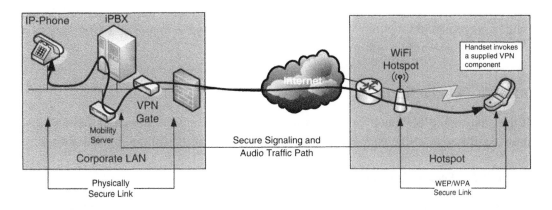

Figure 11.3: Remote UMC hotspot access with native security.

- *Handset/mobility server "native" security.* Such an end-to-end design would require the mobile client and mobility server to collaborate, at the application level, in a vendor-specific security scheme. This could take the form of DES, Triple-DES, TLS, RC4, IPSEC, or AES encryption that was managed independently of any other security policy that might be imposed. Such an architectural approach is the simplest and provides for the simplest remote handset support.

- *Overlaid security.* If not provided natively by the UMC solutions, overlay security solutions may be included in the final system design. These can take the form of support of third-party VPN modules on the mobile handsets that can link with a VPN server hosted in the enterprise. Indeed, many commercial dual-mode handsets have VPN services delivered by the manufacturer. Taking advantage of this security option means configuring the mobile handset and UMC client to interact properly with the VPN services so that when invoked they do not destroy or seriously interrupt the audio traffic.

A technology challenge facing the UMC vendors today is the fact that the processing power of most dual-mode systems is not sufficient to handle real-time RPT processing, codec processing, and encryption services. Even with a 400 MHz processor, there is barely sufficient power to process the packets and manage the quality of the link. The evolution of better mobile devices will take the form of faster processors, dual-core processors, or specialty ASICs in their designs.

A third option for implementing remote security involves access from a site equipped with a VPN server statically linked to its corporate peer. In this model, the mobile handset needs no changes, but by the network design, all traffic leaving the remote site will be encrypted through the VPN tunnel. This is a perfect example of a UMC use of a remote corporate office, but not for a public hotspot.

11.4.2   Media Security

If the end-to-end RTP audio stream is not encrypted by a VPN service, it can be protected by one of the encryption standards described in the following subsections.

11.4.2.1   Secure Real-Time Protocol

SRTP is an extension to the original RTP standard but dictates how to apply AES security to an audio or video stream (IETF RFC3711). In addition, it supports the concepts of message authentication and integrity verification, which means that with SRTP, not only are the frames secure from decryption by an unauthorized entity, but any attempt to capture, modify, and resend the frames can be detected.

In parallel with SRTP, the Secure Real-Time Control Protocol (SRTCP) is also defined. Encryption key management is via external key management protocol. Typically, a *master key* is exchanged and is then used to securely compute subsequent sequence keys.

Many VoIP vendors support SRTP and SRTCP for their IP desk phones and softphone products, but as of the first quarter of 2008, no UMC vendor supported this level of security on a mobile handset.

11.4.2.2   IP-Security

The IPSEC security standard is employed in a number of UMC products currently on the market. Specifically, any 3GPP-UMA conformant product will use this encryption scheme for both signaling and audio streams. The only downside to this approach occurs in running multiple applications on the same mobile device; they all will have to conform to this specific security policy alone.

11.4.3   Security Scope Considerations

For any serious business implementation of a UMC solution, security should be one of the most critical requirements for product selection. Some commercial offerings may have only partial solutions for the security requirements (i.e., secure signaling without secure audio). Others may not have absolute secure control over the possible data paths for UMC application traffic; these limitations must be taken into consideration when making a buy decision. The validity of any UMC solution must be evaluated within the framework of the specific enterprise security policies.

11.5   Balancing Security with QoS

Implementing WLAN voice systems introduces a dichotomy that must be addressed: Imposing security on wireless subsystems is often in conflict with the requirement of low latency for the voice application. To ensure an end-to-end secure voice connection, often there is a penalty to be paid by increased end-to-end latency to the point where the latency is too great and results in unacceptable voice quality (safe, but unacceptable). In the worst-case scenarios, imposing the highest level of security may cause an audio break of 2–3 seconds when simply roaming between two WiFi access points in the same building.

11.6   Multi-AAA Authority Overview

Though most security systems include authentication, authorization, and accounting (AAA) functions, a UMC product may encounter multiple AAA authorities in a live deployment. This is because the UMC application crosses multiple transport authorities and multiple network media authorities, and, of course, the application is an authentication authority itself. In this scenario, a single UMC client may be required to successfully be authenticated with multiple authorities before a single packet of application data is exchanged. Taking this factor into account in the overall application design is important for optimizing the user experience.

11.6.1   UMC Controller

When the UMC client registers with the UMC server, it must pass some authentication test. Additionally, if the server meets certain criteria, the client will *authorize* the server by completing the registration process. This bidirectional authentication is an example of a robust security design and should also include strong end-to-end encryption. This operation can authenticate both user and device.

11.6.2   WLAN Controller

Conforming to the IEEE 802.11 specifications, the WLAN client must also authenticate through its association process. Only after successfully being authenticated can a mobile client receive its security keys and complete the association process of being attached to the LAN. This process only authenticates the device.

11.6.3   VPN Controller

If required by corporate policy or to simply ensure end-to-end security because of an absence of another security mechanism, a successful VPN registration will validate the user but not the device.

11.7 UMC Registration and Security Considerations

Each UMC client will have an associated *network-state*—knowledge of the associated network (WiFi or cellular) and specific network addressing information that is so important for basic operation of such applications. This class of information is usually obtained by the central application control point through a *registration* operation. For 3GPP-UMA applications, registration is performed only once, on initial entry into the network, and because the phone is mimicking a cellular phone. For enterprise-centric UMC solutions and 3GPP-VCC applications, however, registration must be performed each time the device crosses a network boundary. It is the mobility server that must have knowledge of the current network of residence and any specific IP addressing that may be active (see Figure 11.4).

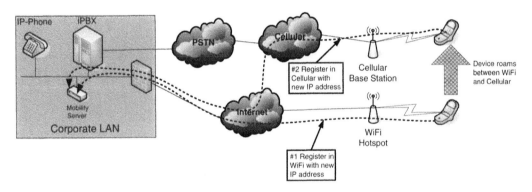

Figure 11.4: Multiregistration requirement.

In such a multiregistration scenario, the UMC application must be designed to successfully complete said registrations while passing all security access requirements upon each network transition. For multiple reasons, a registration operation should be repeated periodically, even if the user remains on only one network. Changes in status and conditions such as IP address lease expiration will also trigger a registration. If the registration design utilizes packet data services on a 2.5G cellular network, the registration operation will be blocked if the user is in an active call. Once the call is completed and the packet service is restored (in a few seconds), the registration may be initiated. In certain cases this scenario will cause a disruption in the ability to make rapid subsequent calls.

CHAPTER 12

PBX Features and Integration

12.1 PBX Features: An Introduction

To be able to make a call from one mobile phone to any other phone in the world, regardless of type of network coverage, is quite a feat, but consumers and businesspeople have been conditioned to expect much more from their mobile phones. Extended features like multiparty calling, voicemail, and call waiting are just a few of the *supplementary* services that today's sophisticated telephony user expects. Business users have even higher expectations for features such as call transfer, conferencing, call park, call hold, auto-attendant, music on hold, and others. Some PBX products support upward of 200 different telephony features; in a UMC context, how will these feature expectations be met? *Should* they be met?

There are two basic approaches to providing UMC PBX-class functions that align with the various UMC architectures. For carrier-centric products, it is the carrier that provides these services. Indeed, today most cellular services provide multicall, voicemail, and three-way conferencing as standard offerings, and a carrier-centric UMC solution replicates cellular features in WiFi coverage. Enterprise-centric UMC products, however, have a more challenging problem in delivering PBX-class features; the solution must integrate to the hosting sites' existing telephony system, which may be complicated by vendor-proprietary implementations. IP-PBX products simplify this problem somewhat, but it is still an uphill challenge. This chapter will describe the various UMC approaches in adding PBX-like features along with the pros and cons of each approach.

12.1.1 Carrier-Centric UMC Solutions

Carrier-sponsored UMC services such as UMA or VCC derive their PBX-class features directly from the base cell phone features. Since such UMC products merely emulate a cellular phone while in WiFi coverage, specific features will be carrier dependent.

To tap into the enterprise market, certain wireless carriers have implemented *pseudo*-PBX functions like abbreviated (or extension) dialing. Leveraging an UMA/UMC solution, such carriers can process a four- or five-digit dial string and, as a separate service, redirect that call to the specific extension managed by the corporate PBX. Once connected to the hosting PBX, the handset can invoke certain PBX functions via *feature codes* entered manually by the user (e.g., "*6" may invoke a *mute*).

12.1.2 Enterprise-Centric UMC Solutions

Enterprise-centric UMC products classically support PBX integration options and may be implemented in two architectural variants:

- *Hosted services.* PBX hosting is off site from the customer, and the mobility capabilities are under the control of a third-party provider.

- *On-site support.* The mobility capabilities are colocated with the premises PBX under the direct management of the company IT organization.

12.1.2.1 Hosted PBX Services

PBX service may be provided without customer premises equipment (CPE) and is often referred to by the Centrex term, which was the original AT&T branding of its hosted PBX model. This solution appeals to some segments of the market because it places less responsibility on the customer for having on-site expertise; basically this is an *outsourced* approach to address the mobility problem (see Figure 12.1). The downsides of this approach are higher cost, possible feature limitation, and lack of direct business control over the cellular users.

Hosted IP-PBX services are becoming more popular, and this business segment is growing to serve the lower end of the enterprise and SMB markets. The problem, however, is that many of these service providers have not evolved technologically enough to add UMC capabilities that complement their IP-PBX services. Identifying

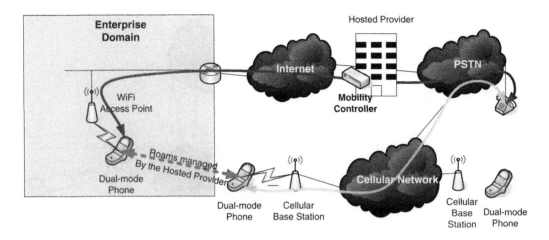

Figure 12.1: Hosted UMC solution.

a full-service UMC-hosted PBX provider might be a regional challenge over the next three to five years and could also have limited selection in the handsets supported.

12.1.2.2   On-Site PBX Services

Most enterprises have elected to purchase their own premises telephony solution as private branch exchanges (PBXs) because of a requirement for control over the telephony features supported by the company, and any mobility solution will have to provide an integration path for this legacy equipment. With a traditional time domain multiplex (TDM) PBX system, the mobile deployment solution involves a third-party gateway to bridge the communication link between the TDM and mobile VoIP solution (see Figure 12.2). It is important to consider balancing the capacity of the gateway with that of the hosting PBX to sustain the proper number of planned maximum calls.

UMC interfacing with a legacy PBX poses many challenges because most are based on a proprietary design, which means that a unique interface solution might be required for each different vendor (or model) that a business possesses. Additional cost may be incurred because of necessary PBX service channel upgrades to support the concurrent mobile call links. UMC solutions that address legacy PBX interface challenges are on the market and available from several reputable vendors. Typically, such solutions are accomplished through a technology and marketing partnership of a gateway and the UMC vendor.

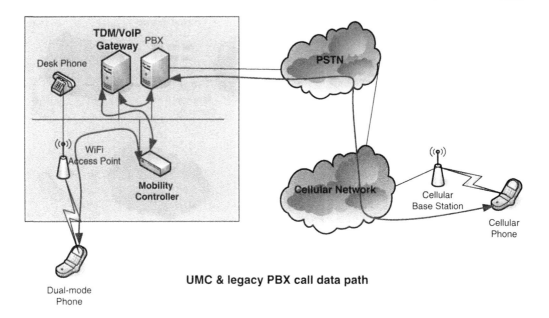

UMC & legacy PBX call data path

Figure 12.2: UMC TDM-PBX interface.

Premises-based UMC products are faced with two PBX interface possibilities that include *trunk* or *line* side configurations; the functional differences between these two approaches are quite large. A premises PBX connects to the PSTN via a trunk interface provided by the central office (CO) of the local phone company. This service is often a multiline channel (24 to 30) bundle with Direct Inward Dialing (DID) supported so that each business phone may have its own, unique 10-digit phone number. It is the responsibility of the PBX to handle the proper routing of the call to the requested phone number deskset (see Figure 12.3).

A PBX trunk interface with a UMC mobility controller will make all the associated mobile handsets appear as foreign phones to those under the direct control of the PBX. Calls from the PSTN intended to be directed to a mobile phone will require special PBX provisioning of a pseudo-DID that is mapped to the foreign phone number via a feature of many PBX systems called *off-system extension* support. Access to voicemail and notification of voicemail (a *message-waiting indicator*, or MWI) may be complex or not supported at all, since these devices are treated like any other foreign phone that is supported through the PSTN. UMC solutions promoted by non-PBX

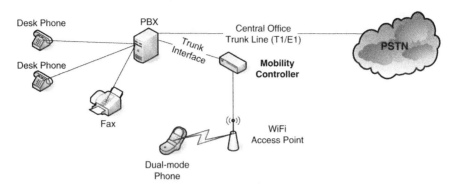

Special handling required for local PBX to support UMC phones:

1. Provision for routing foreign DID numbers mapped to local DID
2. Limited access to PBX hosted voicemail

Figure 12.3: Trunk-side integration example.

vendors may lead their products with trunk-side solutions that will meet approximately 80% of business mobility needs that also support PBX functionality. Single-number "reach" (one number on the business card) is typically achieved by provisioning a call-forward state for a designated DID to reference a cellular number. In such a manner, someone may call an enterprise number but have a cellular phone pick up the call. Access to other desirable features, such as call transfer, conferencing, and voicemail, may be limited because of the trunk interconnect method.

UMC trunk interconnects are the simplest option for mobility vendors, but this requires a mapping of features and capacities matching between the hosting PBX and UMC system. This is especially true with multitrunk configurations where consideration for load balancing and dial-plan redirect are important. Such a state requires proper matching of configurations for both PBX and gateway to ensure proper operation. In planning such a system, a "holistic" view of provisioning must be taken to ensure that the PBX, gateway, and UMC mobility controller configurations are properly matched.

A significantly tighter integration can be achieved if the UMC solution is interfaced through a *line-side* (or *station-side*) connection. This is an interconnect mode where the devices serviced by the mobility controller are configured to be the exact functional equivalent of any other phone serviced by the PBX. Acting as a proxy device controller, the mobility controller emulates a "standard" PBX supported deskset, including the

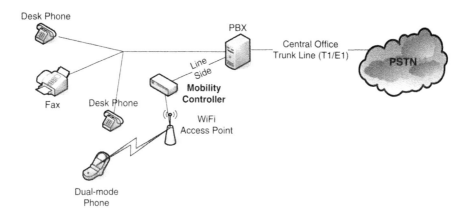

Proprietary handling required for line-side PBX to support UMC phones:

1. UMC handsets have to mimic or proxy the hosting desk phone protocol
2. Exact integration with PBX services possible

Figure 12.4: UMC line-side interface.

appropriate signaling for invocation of special features such as call transfer, call conferencing, call hold, call park, and others (see Figure 12.4). The end-user appeal of this approach is that no special provisioning or configuration changes are needed for the hosting PBX other than defining new subscriber sets to be managed by the PBX. The market-preferred PBX interface option will always be "line-side" due to the minimizing of the configuration management of the PBX and the extended feature capabilities realized through this connection mode. Such integration options will most likely come directly from the PBX vendors as mobile extensions to their standard product lines.

Table 12.1 details the feature disparity between most *trunk-side* and *line-side* implementation options. Because they are more easily implemented and tested, trunk-side offerings will hit the market first. The line-side product options will most likely be associated with iPBX products and not with traditional TDM PBX configurations due to the proprietary nature of most TDM systems.

The market dominance of SIP, as the commercial VoIP of choice, will greatly simplify the way mobility solutions are offered to the market and will eventually accelerate the number of line-side commercial offerings.

Table 12.1: Trunk/line-side feature matrix

Feature	Trunk-side support	Line-side support
Single-number reach	Must configure busy/no answer call forward and unconditional call forward	No PBX configuration changes
Abbreviated/ extension dialing	Supported; mobility controller must be properly configured via dial plan definitions	Supported
Twinning or simultaneous ring	Not easily supported	Supported
Meet-me conference call	Supported	Supported
Ad hoc conference call[1]	Not supported for UMC device	Supported, as other PBX desk sets
Call transfer	Supported – inter-trunk transfer	Supported
Call hold	Supported	Supported
Camp on	Not supported	Supported
Call park/pickup	Not supported	Supported
Message-waiting indication	Not easily supported	Supported
Voicemail pickup	Supported; mobility controller must be configured for access to the hosting voicemail	Supported
Group pickup	Not supported	Supported
Corporate directory display	Not supported	Supported

12.2 PBX Integration Challenges

Businesses that adopt a wireless telephony solution will most likely require integration into their legacy PBX systems. Since a PBX is a rather large investment, it is typically not a consideration to install a new PBX simply to add support for some additional wireless handsets—thus the "adjunct" solution approach. Basically, this product architecture approach is to provide a "gateway" solution that interconnects to the PBX and emulates either an analog or digital handset phone line. The gateway

---

[1] A UMC device can participate in an *ad hoc* conference call if initiated by a line-side deskset.

then acts as a *proxy* for the wireless handsets and accesses them through a defined call control protocol over 802.11 wireless networks. This approach provides a way to preserve the PBX investment while supporting integration of this new wireless solution into an overall business telephony solution.

The gateway approach is predominant in the current market because the major telephony vendors have not fully embraced VoIP as an integrated service provided by the PBX itself (see next section), and customer adoption of VoIP is advancing rapidly. A number of gateway solutions are available on the market, but some consideration should be given to how they integrate into a particular customer environment and the breadth of features supported.

Because the gateway is often a product of a third party (not the prime PBX vendor), integration is provided, but some limitations might exist with these solutions. If your company is investigating a wireless VoIP solution, it is important to understand these possible limitations and how they might impact your particular operation. Some of the limitations found with the currently available gateway products are:

- *Scalability.* Most often, gateways are designed with a limited capacity for handset support. In selecting the capacity of any one gateway solution, judgment should be made by estimating the number of active calls that need to be supported. Of course, multiple gateway units can be configured onto a PBX, but limits in active capacity can impact the total cost and manageability of such a configuration.

- *Limited network capability.* Making a product/vendor selection should involve understanding the network feature requirements. Some solutions place restrictions on how they can be integrated into a multisubnet network. Items such as remote management come to bear with this consideration. Enterprise customers will need to clearly understand these criteria.

- *Limited feature access.* Some cost-effective gateway solutions, however, provide only a limited feature set. Functions such as caller ID, voicemail message indication, and others may be important to an operation and thus need to be considered in making a gateway selection.

Now that we understand some of the considerations of a solution mix, an understanding of the basic PBX connection options is important (see Table 12.2).

Table 12.2: PBX interconnect options

Connection method	Benefit	Limitation	Remarks
Analog	Lowest cost	Feature limited	This option is a "universal" option. All commercially available PBXs support analog connections. This option is usually the cheapest solution but doesn't offer the extended feature set provided by a digital connection.[2]
Digital (ISDN/T1/E1)	Higher device capacity	Feature access limitations	Most commercial PBX products also provide integration solutions via an ISDN or T1/E1 connection. This digital gateway/PBX connection provides a more reliable, feature-rich interface along with a more scalable option. Some vendor-specific features may not be supported on individual PBX systems.
SIP trunk	Relatively low cost	Bandwidth limited based on the base transport link	The "wave" of the future will be using a SIP trunk to provide connections to the private and commercial world, much as digital connections provide today.

Any of the "connection" methods can be suitable but will depend on the specific requirements of the installation site. In evaluating a vendor's solutions, be sure that a clear definition of telephony requirements is provided and reviewed. The good news is that reliable adjunct solutions are available today that provide good voice quality and sufficient features to make it a value add.

12.3 Integrated Solutions

One dynamic that is clearly evident in today's market is that of the telephony vendor's aggressive support for VoIP solutions. Already the sale of IP *seats* has surpassed the sale of analog or digital *seats* for commercial PBX systems. What this means to the enterprise or commercial VoIP buyer is that more solutions with high levels of integration will be commercially available. It takes only a quick literature search to find that all major telephony vendors currently offer (1) hybrid VoIP systems and/or

---

[2] Because PBX vendors and telephony service providers want their customers to move to digital connections, they will often artificially price analog above digital to persuade the customer to move "up."

(2) pure VoIP solutions as part of their product lines. This market megatrend results in one major impact on the resultant configurations: *no gateway requirement*.

The "native" VoIP support provided by these products affords solutions that:

- Reduce the number of vendors involved in the deployment

- Are more cost effective, incrementally lowering the cost per handset

- Typically extend the functionality and reliability of the system

- Potentially support VoIP toll-bypass capability

As "native" VoIP solutions make their way into the market place, industry and enterprise businesses will begin a wholesale adoption of this technology simply because of its wide industry support and cost benefit over the older TDM systems. In investigating a VoIP solution, it is important to work with the telephony vendor and understand how its wireless VoIP offerings (if any) integrate into the whole solution being offered. As the technology finally matures into the market, the compelling ROI and lower TCO will make VoIP and wireless VoIP a success.

12.4   PBX and iPBX Interoperability Future

There are good, solid solutions in today's market whereby a business may deploy a UMC product and leverage its existing PBX investment. Contacting the UMC manufacturers and learning about the connectivity options offered is the best method for understanding the "system" option. Integrating a UMC solution into a legacy PBX system may involve interfacing with multiple vendors, and the UMC vendor must be knowledgeable of the options. Learning the configuration options will be central in determining the proper final product configuration.

If the current corporate telephony vendor offers a VoIP solution (VoIP desksets and/or PDA software), it may also offer a UMC solution that is compatible. Whether your final solution is a hybrid-PBX or a full iPBX, check with the telephony vendor first regarding any UMC solutions it supports. There are also third-party UMC products that should be investigated.

The final word regarding planning for deploying a UMC solution is that no matter what vendor(s) you purchase the components from, it is very important that the underlying installed RF 802.11 infrastructure be certified for voice coverage and that it provide that key voice prioritization feature.

12.5 UMC Feature Supplementary Service Requirements

UMC solutions typically do not natively provide PBX services; these must be derived from some complementary system component. To the user, however, the mobility and the features provided must be unified and seamless; the PBX features are often presumed to be present and the mobility was an add-on. To support these features, any UMC application must account for these features in its base protocol and GUI designs. The way such features are propagated across various wireless networks and the way they are presented through the user interface become critical aspects of a UMC applications success. The following sections describe how UMC systems may provide support for these important features.

12.5.1 Call Hold/Mute

Call mute and call hold are often confused with each other and presumed to be the same feature, but from an architectural perspective they are vastly different. Mute is a client-side capability that when enabled only halts the transmission of audio from that device. The muting party continues to hear the other party, but the peer hears nothing. Call hold is a PBX-managed feature that requires coordination among the UMC mobile client, UMC mobility service, and the PBX. When a call is placed on hold, signaling must take place from the client, through the UMC service, to the PBX, which takes over call control and invokes music on hold (MOH) for the party being held. At that point, the proactive client has no audio connection with the held party. Call hold is usually invoked pending a call transfer or a long period of side consultation that does not involve that person.

UMC call mute is a simple feature to implement, but call hold may be vendor-specific in its implementation with commercially available PBX solutions.

12.5.2 Call Transfer (Attended or Unattended)

In a business context, the ability to transfer a call to the right decision maker is very important—one of the top five critical PBX features required by business-focused UMC solutions. Two types of transfers can be supported:

- *Unattended (unsupervised).* The active party informs the other that they will transfer the call and the new connect will be made without exchange of dialogue with either party. Sometimes this is called a *blind transfer*.

- *Attended (supervised).* In this case, the active party informs the peer that they will attempt a transfer and temporarily place the person on hold while the active party discusses the transfer with the intended receiving party. If the intended party is available and accepting, the active transfer party completes the connection between the new party and the original party. If the intended party is busy or not available, the active party can return to the call with the original party to discuss other options.

The first form of transfer is fairly straightforward to implement with any PBX connection. The latter form, however, may involve vendor-specific features implemented by the UMC solution provider to fully support. Attended transfer is the preferred transfer feature for business because you can always be assured by the knowledge that the call will or will not be transferred successfully. If the transfer target party cannot receive the call, the originating partner can direct the call to their voicemail.

12.5.3 Call Conference

After transfer, the most important PBX feature is conferencing. The ability to bring geographically dispersed individuals together for a call where all can have full, multiparty dialogue about critical topics is vital to business. PBXs and service providers have long provided such features, but this kind of feature has been relatively limited for cellular phone users. Cellular providers have been able to support three-party conferencing, but this has its limitations when trying to include larger groups for quick decision making on important business problems.

When a UMC solution is feature linked to a PBX, it can take on the behaviors of a mobile desk phone; in this mode, it has all the conferencing power afforded the desk phone. Not only can it participate in an ad hoc conference that may be initiated by a peer, it can be the initiator of such conferences just as though they were sitting at their desks. This capability is only found with the enterprise-centric UMC solutions and is absent in the carrier-centric solutions.

Depending on the specifics of a UMC implementation, there may be limitations in support of conferencing. For example, if a UMC client initiates the call, channel demand on the hosting PBX may be exceeded due to the call path architecture implemented. There may also be PBX vendor-specific requirements that must be addressed by the UMC vendor on a case-by-case basis.

12.5.4 Call Waiting

When you are on a call, it is important to know if a second call to you has been made. Indication of an incoming call occurs through call-waiting notification and may be audible and/or visual in the form of display of the call-waiting caller ID. Depending on the sophistication of the UMC client, the user may be able to "swap" calls, auto-holding the current call and taking the new incoming call. Once in this mode, the user should have the ability to toggle back and forth between the two calls on demand. This kind of behavior has been common with cellular phones and is expected with new mobile solutions. If the user chooses to ignore the incoming call, after a period of time, this call should be directed to voicemail for retrieval at a later time.

An additional value-added feature with support of two active calls is the ability to converge these individual calls into a single conference call at will. This feature, however, is not supported in most current UMC solutions.

Though it is technically possible to support more than two active calls, a third incoming call is typically autorouted to voicemail. It is difficult for individuals to juggle more than two calls at one time; such limitations have become accepted as the norm for feature support.

12.5.5 Voicemail and Message-Waiting Indication

12.5.5.1 Voicemail

Support for PBX-hosted voicemail and notification of message-waiting indication (MWI) are some of the greater challenges for enterprise-centric UMC solutions. For carrier-centric UMC products, this is a fairly simple problem in that there is a single voicemail object to manage. For enterprise-centric UMC solutions, there is a possibility of two voicemail subsystems (or more) to manage:

- Carrier voicemail

- Business/PBX-based voicemail

Because of the enterprise-centric UMC integration with the company PBX, there is virtually no need for continuance of the carrier voicemail server that is typically provided with the carrier's service-level agreement. If a user intends to continue using a public carrier phone number in addition to the assigned corporate UMC phone number, there is a rationale for retaining both these systems. To simplify access of voicemail for the user, there is typically a menu function that is invoked for calling into the voicemail system for audio-level access. If both carrier and PBX voicemail systems need to be supported, this poses an additional challenge to the features exposed in the UMC client.

A new format for voicemail access is emerging: *visual voicemail*. Through this mode, active voicemails are not accessed as audio streams but rather through presentation of a displayed list of members. The caller ID, phone number, time of call, and length of message are usually displayed for each voicemail in the list. The user then has the option of scrolling down the list and selecting one to listen to. The user may choose to delete items in the list without listening to them. Such a presentation mode makes remote management of voicemail much simpler than mere sequential listening to each voicemail in received order.

Support of visual voicemail is a real challenge for UMC vendors because the UMC application is not tightly coupled with the hosting PBX voicemail systems and there may be no simple PBX services to build such a list for display. Support for such an extended feature will most likely be supported on a PBX vendor-by-vendor basis where the proper interface services are available. Without visual voicemail, the user can still make a call into the voicemail system and retrieve it as audio output.

12.5.5.2 Message-Waiting Indication

If a voicemail is recorded while you are on the phone or the phone is turned off, it is important to know that a message is waiting to be retrieved. Even with visual voicemail, some indication that a new message is available should appear on the main UMC screen. With carrier-centric UMC products, this is not a major issue since they are merely emulating a standard cellular phone and there is only one voicemail repository. For enterprise-centric UMC products, support of MWI can be a very tough challenge. Exactly how MWI gets supported, whether in real time or batch, can be a key value-add for a UMC solution.

Whether or not MWI is supported by UMC products may be determined, in part, by the way it is interfaced with the PBX. If it is interfaced as a line-side connection, most likely this indication will be signaled when the PBX gets the message and the UMC client must accommodate displaying that state in the GUI. If the connection is trunk side, there will be no simple method for supporting MWI, because the mobile devices are treated like any other device that is addressable in the PSTN. Support for this feature, if at all, will be on a PBX vendor-by-vendor basis.

12.6 Network-Specific Implementation Challenges

UMC solutions that are focused on the enterprise market are not without their limitations when considering all possible mobile use models. Because in such systems call signaling and audio traffic homerun back to the hosting PBX/mobility server complex, the call/channel ratio on the PBX is doubled (see Figure 12.5). When a call between UMC phones is made from outside the corporate WLAN coverage, one audio channel is required for the link to the initiating phone and one for the receiving phone. The looping back through the PBX is often termed *tromboning* or *hair pinning* and is an artifact of the base architecture; such a requirement must be taken into consideration for capacity planning.

Beyond the problem of increasing the number of PBX ports, this problem aggravates cost-saving efforts when calls are made internationally over mobile networks or across carrier networks that do not have reciprocating roaming agreements (see Figure 12.6). Without a UMC design that considers international cellular cost models and tunneling between mobility servers, an international call can end up with a cost burden for four call legs between two individuals!

Some UMC solutions will block automatic roaming to foreign carrier networks and prompt the user prior to a roam with an indication of a cost impact, but the optimum UMC least-cost international routing design is to construct "networks" of regionally placed mobility servers that communicate with each other and act an internal traffic routing services that bypass the cellular networks (see Figure 12.7). In this scenario,

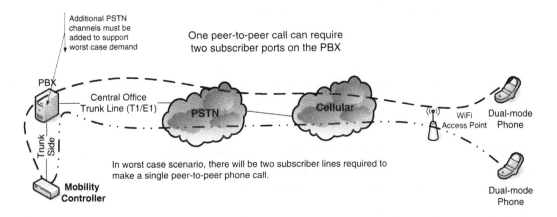

Figure 12.5: Strained capacity on PBX.

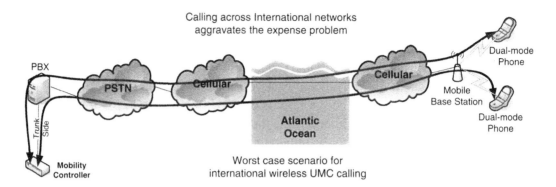

Figure 12.6: Worst-case international calling.

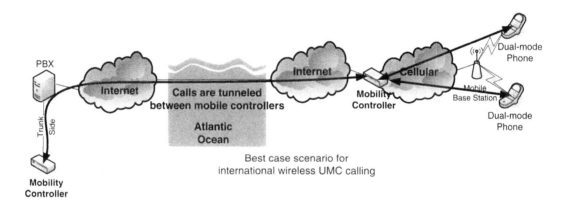

Figure 12.7: Best-case-scenario international calling.

placing a call on a dual-mode phone in a foreign location would ideally result in a discovery operation to identify a regionally located mobility service and to set up a call that would be tunneled through the local mobility service to the destination mobility server over the Internet. In this manner, the only cellular calls to be charged would be local calls rather than international calls.

The problem with this approach is that no current commercially available UMC vendor can support this configuration, and it would also require a Global 2000-class company to design, deploy, and manage such a network structure.

Of course, with any international call, it is preferential to make it as a WiFi call from the remote office or appropriate hotspot.

For large enterprises interested in deploying a UMC solution, another hurdle to address is capacity. It is possible to conceive of an entire corporation going "mobile" and doing away with desk phones altogether. In such scenarios, the number of clients to support would easily be in the thousands. This solution scope demands a new level of system features focused on (1) load balancing, (2) high availability (HA), and (3) single point of management. Naturally, a high user capacity would be accompanied by a higher solution cost, but such mobility solutions would be implemented as multiple servers (colocated or not colocated) with dynamic network self-healing in case of component failure and with load balancing being performed automatically. Such solutions require longer-term development cycles and might not appear on the market until 2009 or 2010.

12.7 PSTN Interconnect Options

At some point in the distant future, the PSTN might not exist, but until that time any successful UMC will provide for some interconnect to the PSTN. From a societal perspective, it may take several generations before wire-to-the-house is no longer the standard telephony service. Given this table stakes requirement, the following subsections outline some basic information about UMC/PSTN connectivity considerations.

12.7.1 T1/E1/J1/PRI

Most businesses own their PBXs and have chosen a digital CO connection option. In North America, this is typically a T1 line, which supports up to 24 simultaneous calls over that resource. In Europe it is an E1 line that provides for 30 concurrent calls. Whatever PSTN connectivity is available, the customer needs to decide the worst-case usage model they want to provide their associates. For example, a company of 50 people might only install a single T1 line on the assumption that no more than 50% of the employees will need to be on a call at any one time. Such a model is called "over subscription," and there is some level of risk that a call may be blocked when all the channels are in use. At a cost of hundreds (or thousands) of dollars per month per T1, the level of oversubscription risk is a subjective business decision.

In the case where multiple T1 connects are available, it is important to understand how the UMC solution interconnects with such configurations and what impact it may have on these resources (see Section 12.6).

12.7.2   Analog

Enterprise analog connectivity to the CO is almost nonexistent in the 21st century. Many homes still have this class of service, but serious businesses will prefer a digital connection to the service provider. Where analog may have a play is on any station-side connection options. Many PBX vendors still support an analog connection option to desktop phones, and a UMC solution may take advantage of this option because of sunk costs. To take advantage of this situation, an analog⇔SIP gateway will be required. These gateway products are commercially available in four to 32 port versions.

Use of a UMC analog interconnect should be seriously questioned because of loss of key digital features and lack of future support. Features such as caller ID, call waiting, and MWI are typically not supported on an analog service. Additionally, having to add new analog ports on a PBX is often more expensive than adding digital ports, because the PBX vendors are trying to phase out support for this older technology and use pricing strategies to drive customers to digital connections.

12.7.3   SIP Trunks

A SIP trunk interface from a service provider is the simplest and most often the least expensive connection for a UMC application. These are the IP logical equivalents of a T1 or E1 line (although not in total capacity). With such options, the question must be addressed as to how calls get routed to the PSTN. With a SIP trunk connection, this is presumed to be provided by the provider. In such scenarios, feature and cost tradeoffs must be made because direct access to PBX supplementary features may not be available. The full availability of desired features must be evaluated based on the specifics of the proposed configuration.

CHAPTER 13

UMC Management and Statute Conformance Considerations

Once the technological hurdles have been addressed and a commercial UMC solution is made available to the general market, a question that arises is, "How do I manage these devices?" This issue is generally not a consumer UMC play, because the individual user typically has no major management concerns other than ensuring that they have physical control of the device. Business-targeted UMC solutions are another matter altogether. Deployment of a UMC system within a business requires capital expenditures with the expectation that resources will be managed just like any other networking or telephony resource to maximize the return on investment (ROI).

In the context of business, a UMC solution must address a number of business concerns:

- How is the mobile device managed as far as ensuring secure access to the company network and company data?

- Is it possible to monitor the way this mobile device is used—for personal or company purposes? Both?

- Is it possible to remotely manage the UMC application version and configurations?

- If there are user problems, what facilities are available to troubleshoot the problem?

- What happens if the mobile device is lost or stolen? Can it be disabled remotely?

- Are there laws, statutes, and specific market requirements that apply?

The answers to these questions may play a key role in making a buy decision for a business seriously looking for a highly mobile communications solution. In a business context, the device and investment responsibilities are shifted from the actual user of the mobile device, as in a consumer device, to the corporate owner of the device; the requirements for consumer-based and enterprise-based products are radically different. Additionally, the merging of technologies results in new requirements for management and reporting that have not existed in previous products.

13.1   AAA Management Requirements

Chapter 11 touched on the security aspects of AAA: authentication, authorization, and accounting. One great feature afforded by a UMC solution is a singular login, which now allows corporate IT to manage access to cellular phones where previously this was not possible. The management and reporting aspect of the accounting capability manifests itself in the requirement to log, report, and alert a management entity of any failures. Such failures might be indicative of unlawful intrusion attempt, component failure, or simple user error. Retention of such transaction histories also becomes important in terms of being able to analyze trends or error patterns.

Beyond reporting errors or suspicious conditions, UMC systems demand a new level of features that are important to a business. Often the reason for purchase of a UMC system is cost control over the cellular resources. In such cases, an augmentation of standard telephony reporting of the call detail records[1] (CDR) is required. Mobile detail records (MDR) document network utilization for an associated call through reporting of WiFi and cellular minutes used. Aggregate reporting of such statistics helps a business assess how effectively its UMC system is being used and what changes might be suggested to improve the general usage models or carrier SLAs.

CDR reporting from a PBX is *after-the-fact* call reporting. With a UMC solution, *real-time* reporting is possible to enforce business policies; this is especially true regarding regulation of cellular phone usage. Standard cell phone services don't provide real-time management features that would allow a business to monitor cellular usage and impose corporate policies on the cell phones as they do on desk phones. With a

---

[1] Who made the call, to whom was it made, duration of the call, and so on.

UMC solution, the realization of this level of monitoring is possible, to manage who is being called and how many cellular minutes are being used. Network quality assessments can also be derived from such histories, allowing corporate IT to better plan and manage its networks.

13.2 Physical Security Concerns

What about security control over the physical mobile device? What if someone steals the unit and knows the user secure access information? In such cases, risk of uncontrolled telephony usage would be possible in addition to loss of valuable corporate contact information. Not only could someone make phone calls that would be charged to the corporation, but someone could extract from the device valuable customer and associate information that can be used competitively.

Once the UMC management authority was alerted that a unit was stolen, the first action would be to simply disable the device itself. Depending on the flexibility of the UMC management solution, the order of action would be: (1) disable the device followed by (2) disabling the associated user. Loss of user ID information is serious because this information can be used to access multiple devices from the same company. Disabling the device is also important in case the thief has multiple user name/password sets.

Disabling the user and the device are merely the initial actions to take. Even if the thief can't log in and make calls through the corporate system, he does have access to handset-resident information. Having the ability to remotely "kill" the handset is very important, but this presumes a valid network link. Concurrent with any termination of a device/user configuration, a method of assigning a "kill" to that device is important. In the case where someone attempts to use that phone at any time, the login will fail *and* the phone information can be deleted. Unfortunately, there is no absolute protection against someone stealing data from a phone if they do not attempt a login. The only defense against this possibility is to (1) encrypt the data resident on the phone and (2) require a login to the physical device itself.

The one other concern about a stolen UMC device would be the SIM associated with the GSM services. This SIM can be removed and placed in another GSM device and used without hindrance. The only defense for a corporation would be to notify the hosting carrier and have the carrier deactivate the SIM.

13.3   UMC Policy Management

In an ideal world, a robust UMC solution would come with a robust policy management system and integrate into the user's Personal Information Manager, or PIM (see Figure 13.1). In this world, the system manager could dictate such policies as:

- Who the user was permitted to call (local calls only, certain area codes, 800 numbers only, company associates only ... ?)

- What carrier services were acceptable from a service and pricing standpoint?

- Presumptive presence based on the user PIM; no phone calls during a schedule meeting or automatic do-not-disturb based on workday/weekday hours.

- What is the best communication method based on the user's geographic presence (i.e., what is the delta of the time zones between the two parties)?

- What are the hours for enabling the corporate cellular service and what hours for disabling the corporate number?

Access to other corporate applications would be extended across the wireless network domains to eventually include the whole worker's data/voice job environment. In such a

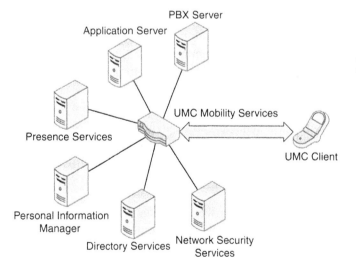

PBX Server
Application Server
Presence Services
UMC Mobility Services
UMC Client
Personal Information Manager
Directory Services
Network Security Services

A completely integrated UMC solution would provide services to:

- Manage and report who called or was called by the UMC device.
- What networks (cellular and WiFi) were permissible for use
- What was the most reliable network service to meet the application need.
- When and what kind of media would be appropriate for communicating with a user.

Figure 13.1: Ultimate UMC architecture.

configuration it would also be important to extend certain controls to the mobile user in manipulating the contact media preferences (phone or text messaging), presence state, and other personal profile characteristics.

Corporate directory services would facilitate mobile access of virtually any resource within the corporate LAN. The corporate directory would also be converged with the user's personal phone directory to present the simplest dialing interface.

Because an enterprise-centric UMC dual-mode phone is a schizophrenic (having two different phone numbers for two different uses), the UMC application must accommodate modes where the phone can be used as a "business" phone and modes where personal phone calls may be made bypassing the corporate policy enforcement.

Some management requirements may impose a *record* feature on a UMC system. Certain legal statutes such as Sarbanes-Oxley require logging of written and verbal transactions for later legal validation reasons. Other uses for recording conversations involve those between a stockbroker and her customers or between a lawyer and client. Such features will be required of a sophisticated UMC solution.

The number of discrete policies is virtually unlimited. Such capabilities would expand to become an entire subsystem unto itself. Indeed, in the enterprise-centric UMC solutions, this will be an area of value-added growth over the next two to three years.

13.4 CALEA/Lawful Intercept Support

In late 1994, the U.S. Congress enacted the Communications Assistance for Law Enforcement Act (CALEA). The law defines the statutory obligation of telecommunications carriers to assist law enforcement in executing electronic surveillance pursuant to court order or other lawful authorization. It is through CALEA that state and federal officers are permitted to "wiretap" an individual or groups of individual phone lines under court sanction. Of course, with the advent of VoIP, new consideration must now be given to define how CALEA support is implemented in this context.

Initially, CALEA was carrier (wireless and wireline) focused; compelling telecommunications carriers to assist law enforcement in executing electronic surveillance pursuant to court order. Because these large carriers had nationwide network presences with convenient points of intercept, supporting CALEA was not a major technical problem.

Currently there is no directive in the statute regarding CALEA support for WLAN or VoIP implementations, but such support should be expected in the future, since companies (or cartels) could set up their own private VoIP networks for communication across state and international boundaries through the Internet. When this situation arises, there will be several technical and logistical hurdles to overcome in implementing CALEA support in a UMC context:

- *Who are the parties in the call?* Any node on any network with any IP address can initiate a call with any other network node. The path of the signaling and audio stream may be undetermined and variable, making it difficult to pinpoint a point of intercept. Additionally, the call initiator and receiver may have nonstatic IP addresses that change during the course of the call and make identification of the session difficult.

- *How can we trace the call when roaming off network?* Tracing an IP-based RTP stream is possible, but a WiFi⇔cellular roam crosses network boundaries and would require a different and coincidental intercept point within the hosting network.

- *What about encrypted audio?* The encryption must be provided to the lawful intercept agency, and a UMC call that roams across multiple networks might have multiple authentication and encryption schemes applied to the single call session.

- *International considerations?* What about multiple national communication laws? Since the Internet is international and the geographic locations of nodes may be difficult to determine (if not impossible), identifying source and destination nodes will be difficult, and the signaling and audio streams may cross multiple national boundaries where different laws apply.

The one obvious solution to this problem would be to have the UMC mobility server become the point of control for the CALEA intercept. In this case, an enterprise might act as a "carrier" to provide any call-recording capabilities. How are such entities identified in a free enterprise society? How would a federal agent know what building or associate to approach to enable a CALEA intercept? All these questions will be answered in the future with application of new statutes to help manage and address these questions.

13.5 HIPAA: Healthcare Considerations

Discussion of implementing UMC solutions in a healthcare environment brings up questions about compliance with the Health Insurance Portability and Accountability Act of 1996 (HIPAA). HIPAA was enacted to protect the health information of individuals by controlling with whom and when such information could be shared. The statute requires implementation of safeguards and procedures for administrative, technical, and physical management to protect the confidentiality of patient data under control of the healthcare system. For any person to receive personal health information on a patient, they must be preauthorized by the patient or patient's lawful representative. This usually takes the form of a patient signing a release form that is associated with the immediate health issue. Once in place, this information can be shared with the designated family and healthcare providers on an as-needed basis.

There is nothing in the current HIPAA statute regarding requirements over WiFi-VoIP or cellular calls that pertain to transfer of personal healthcare information, but some questions arise about authentication and transmission of such information to a mobile/remote device:

- *Is there an authentication process in place that positively identifies the recipient of the information?* A nurse can call a doctor's cell phone, but there is no validation of the person answering the phone other than subjective voice recognition. If the nurse leaves a voicemail, there is no guarantee of who picks up that call and listens to the message.

- *Is there adequate control of the audio playback on a healthcare call?* There's no absolute guarantee that individuals on the periphery of a public area phone call might overhear personal healthcare information and violate the intent of HIPAA. There are commercial products whereby voice communication is made to a voice-badge device that uses a speakerphone for audio. In such cases, there is the possibility of individuals overhearing sensitive information they were not authorized to receive. For both voice badges and phones, it is best if the call participant move to an area where they can minimize the risk of being overheard. Also, speaking a patient's name while on a call in a public area should be avoided.

- *Is the transmission to a mobile device secure/encrypted communications?* Since mobile information might traverse the Internet via a VoIP link and be exposed to potential eavesdropping, ensuring that there is sufficient link security to prevent unlawful intercept is important.

- *Is there an audit trail for all mobile calls where HIPAA may apply?* If there is ever a legal question about conforming to HIPAA privacy intent, are there full audit capabilities in the UMC solution to address such concerns. Besides having CDR reporting, the ability to record such conversations for future playback would be critical.

Today there are no commercially available UMC products claiming conformance with HIPAA, but in deploying UMC in hospitals and clinics, consideration for these concerns must be made. There is no formal HIPAA certification program, merely an assessment of a site's compliance with the statute's written requirements and intent. It is up to the UMC system owner to create and enforce policies and procedures that comply with the letter and intent of HIPAA.

13.6   E911/Emergency Response Support

Support for emergency 911 calls with a UMC solution poses some interesting problems. To be able to properly respond to an emergency situation initiated from a 911 call, knowledge of the location of the emergency is paramount. The cellular system complies with this requirement by directing the call to a regional law enforcement center or a public safety answering point (PSAP) where the audio link is made to obtain the emergency information. In making the connection with the PSAP, the geographic location is also provided (see Figure 13.2). This can be GPS location information or

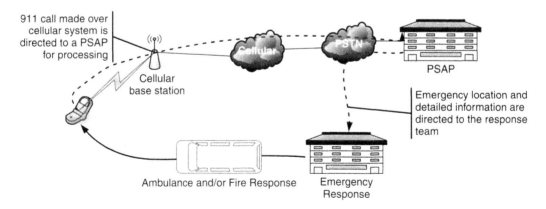

Figure 13.2: 911 emergency response cellular model.

knowledge of the current cellular tower associated with the phone. Once the location information and additional incident information are obtained from the caller, the emergency response teams can be dispatched.

With traditional 911 from a static landline (nonmobile) phone, the location was already known. A 911 call from a home number would presume the street address associated with that phone number, and the emergency response team would be dispatched to that address. Having access to the location (latitude/longitude or street address) is the lynchpin piece of information in making 911 work. Without it, it would be virtually useless. Emergency calls from UMC devices must be deployed in a manner that ensures the safety of the individual making the call.

With the advent of VoIP, new challenges were posed to support 911. With a static, IP-Centrex-like service where the user obtains a VoIP service over the Internet, the procedure was for the user to fill out a *location declaration* form when signing up for the service. In this way, when the VoIP provider received a 911 call, they had the location information to pass on to the emergency response teams (see Figure 13.3). Most home or business VoIP providers now follow this practice.

Add the mobility element to VoIP, and the 911 support problem becomes more challenging. Since a WiFi-attached UMC handset can be operational in *any* valid WiFi/ Internet connection, there is no way to automatically infer the location coordinates. Consider an enterprise-centric UMC user visiting a coffee shop hotspot in London who dials 911. The intercepting ER agents would most likely be in their home office somewhere in the United States, but there would be no way to effectively direct an emergency response team to the user due to a total lack of location information. Even if

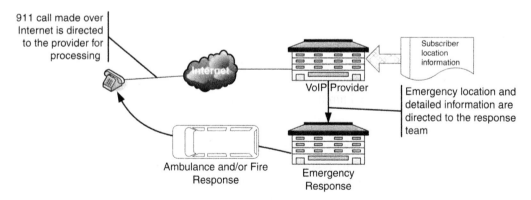

Figure 13.3: Emergency response static VoIP model.

the location information was available, notifying the proper location authorities might be problematic due to a lack of international emergency directory access.

Because a UMC handset is dual-mode, most UMC solution providers rely on cellular services for processing emergency calls. The scenario may be one in which the WiFi signal is strong and of high quality, but when processing a 911 call, the call will be diverted to the cellular network. This is to ensure that the location information can be provided to the emergency response teams.

The mobility of a UMC application also poses additional usability considerations and challenges for the user concerning emergency call handling. The first challenge is the fact that there is no single worldwide emergency phone number. Though 911 is the standard emergency number in North America, the European standard is 112. In general, there is an agreement with the GSM service providers that if a user dials 112 on any GSM network in the world, the call will be forwarded to the appropriate regional emergency response center. There are notable exceptions to this rule, with restrictions as to phone configuration and/or applicable national communication laws. Due to the urgency of an emergency call, most providers allow such calls to be made even if the keyboard is locked and there is no SIM in the phone. This is, however, not true in all cases. Depending on the servicing carrier, there could be a requirement for an account with that provider or the presence of a SIM, or the 112 call may be directed to fire or police instead of a healthcare emergency team. It is *very* important to understand the behavior of an emergency request related to the country or state locale of the call. Such idiosyncrasies of the emergency response calls are not the fault of the UMC vendor but rather quirks in the regional regulations and services supporting emergency services.

Alternative emergency call services may be considered by enterprises for handling "on-campus" problems. In many businesses, office or plant emergencies may be handled by the company safety team or an outsourced safety service. Supporting this emergency response model will be somewhat simpler for a UMC provider taking advantage of the WiFi/Ethernet connection while within the confines of the company LAN (see Figure 13.4).

With this model, the UMC system must support a feature that (1) recognizes that the user is *on-campus* and (2) converts a 911 call into a configured local extension or forwarded call. Support of this feature will require an advanced UMC solution that is not currently on the market but that may be offered in the future. If the UMC user left the office network environment, a 911 call would be processed in the same manager as any cellular 911 call.

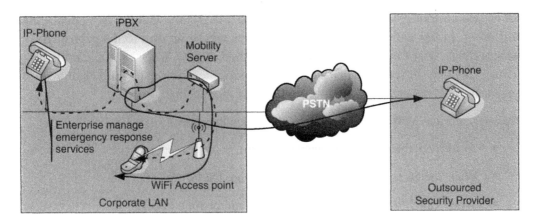

Figure 13.4: Emergency response local service model.

13.7   Securities Exchange Commission Considerations

Some business applications require that calls be recorded for legal purposes; such is the case with stock traders. Any call a stock trader makes with a client that discusses stock availability or pricing must be recorded. Each call must be recorded and catalogued for future inspection if there is a question regarding the agreement or stock trade details. This is for the protection of both stock trader and client. Not all enterprise-centric UMC solutions have the capability to support such a feature.

13.8   Presence Management

Presence—knowing whether someone is available or not regarding a critical call—is an ancillary feature for UMC and can be a real asset. Considered one of the *management* aspects of a UMC product, presence information can be a powerful value-added feature for the system. Allowing the individual user to manage her presence state is very powerful. Sophisticated implementation will report not only that a user is "present" but will also indicate what kinds of communication methods are available at that time. For example, when a user is in a meeting, it is convenient to set the user's presence to *receive text only* for instant messaging. In this mode, the fact that phone calls will be blocked for a period of time should not discourage people from calling, because calls will be directed to voicemail. For some situations, a user may want to set her state to *do not disturb* and turn off any communication mode.

As presence services become more sophisticated, integration with the individual PIM will occur. In this case, certain automatic policies can be applied to the published presence state based on personal schedule and time of day. When *on vacation* is the status, calls may be dynamically redirected to associates taking the individual's responsibilities rather than simply being directed to voicemail.

Presence services will evolve to answer the following questions:

- Am I available to be contacted? (yes/no)

- What media is best to contact me? (phone, IM, SMS, other)

- When can I be contacted? (time of day and/or time zone)

- What is the disposition of unanswered messages? (voicemail or IM history)

13.9   Cost-Control Policies

One of the siren songs of VoIP was that it promised free long distance calling. Indeed, in a controlled enterprise network environment or with use of popular VoIP applications, individuals may make ad hoc, peer-to-peer phone calls at no cost. In practice, however, VoIP cannot completely fulfill this promise due to a number of issues that are critical to businesses and individual consumers. Considering leveraging WiFi/Internet connectivity with a UMC application for telephony cost control further compounds the challenge of meeting the "free" goal, and *least-cost routing* becomes an important feature for both UMC solutions to minimize telecommunication costs on a call-by-call basis.

For a carrier-UMC solution, the cost benefit to the consumer is a potential of reduced per-minute cost on calls originated from WiFi. Early carrier FMC products charge a flat monthly rate (approximately $20) for calls originated from a sanctioned WiFi hotspot. The problem was that such sanctioned hotspots were limited to home or a short list of commercial establishments (i.e., Starbucks or McDonalds). This meant that to take advantage of the "WiFi minutes" rate, the user was restricted to home or a coffee shop—not a very realistic scenario for a mobile individual.

For an enterprise-UMC solution, the options for implementing a richer least-cost routing solution are much greater. Since the base WiFi service is typically an "on-campus" configuration for the enterprise, a UMC solution should prefer a WiFi connection over a cellular connection. In this scenario, calls made between associates *on-campus* would all be "free," just like intercompany calls from a deskset.

Making WiFi calls from a company-sanctioned hotspot (home, hotel, airport, remote office) would also be free for calls to fellow associates reachable over the IP network. Handing off calls to the cellular network would be an exceptional case where the WiFi was weak or nonexistent. In a case where the caller maximized his time within some WiFi network coverage, the magnitude of required cellular minutes would be reduced. Some analysts have stated that upward of 40% of all cellular calls made in an enterprise are made in sight of an office desk phone.

Assuming that a monthly per-user reduction in the cellular minutes could be achieved, several cost control options would be open to the enterprise. Depending on the flexibility of a business's SLA with its carrier, a reduction in the size of the *minute pool* might be renegotiated. If this was not possible, the opportunity of mobilizing more company employees could be investigated for the same budgeted amount. In an organization where there were large numbers of international travelers, significant cost savings could be achieved through a UMC deployment. Each case, however, will be different, and the cost analysis for the ROI must be explored in detail.

13.9.1 Cost-Control Minutiae

With a UMC implementation, the way handoff decisions are made can seriously affect cellular usage costs. In a case where a user is positioned between a strong WiFi signal and a cellular signal, the dual-mode phone should prefer connections through the WiFi link. However, if the user moves throughout a facility, variances in WiFi coverage may be encountered, and the unit is forced to roam to cellular to sustain the call. If an individual is moving through a building with WiFi coverage, it is very likely that within a few seconds after experiencing a weak WiFi signal, the handset will enter an area of strong coverage. In this scenario, it is possible for the UMC handset to transition between WiFi and cellular networks with a relatively high (subminute) frequency. This "ping-pong" effect (see Figure 13.5) will drive cellular minute usage through the ceiling. A handover between wireless networks takes only 10–15 seconds, but each time a cell phone connects to the cellular network, one minute of usage is charged. It is highly possible, then, to incur a two- to three-minute charge within just 60 seconds. This may not have a major per-person cost impact, but with aggregate effects in large organizations, this can become a serious problem. A sophisticated enterprise-UMC application must be aware of this condition and compensate to minimize loss of effective cellular minutes. A guideline for cost management could be, "If you hand over to cellular, remain on cellular for at least one minute."

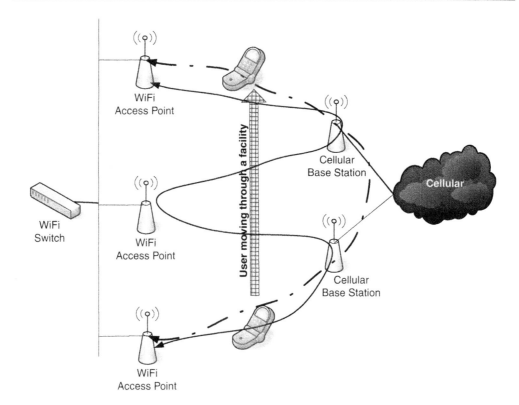

Figure 13.5: UMC ping-pong.

13.9.2 Trading Features for Cost

If cost is a major driving consideration even for cellular management, one control option open for UMC support is direct cellular-to-cellular calling. Just as peer calling within a wireless carrier is now free, making such calls under control of a UMC system could still take advantage of the zero cost. Supporting this feature variance, the UMC system would support definition of call policies for individuals or groups, whereby cost overrides features. Such an implementation would still log the fact that the calls were made and would fall under policies governing who could be called, but the call would be placed as a direct cellular call. This kind of solution has several repercussions:

- *Loss of UMC managed features.* Because the call management is no longer centered in the UMC server, no PBX-centered function can be supported for this call. There's no support for extension dialing, call transfer, hold, or conferencing. Also, there may be caller ID confusion resulting from the receiver getting a different caller ID than normally assigned to the caller.

- *Requires additional contact knowledge.* Free calls are usually realized only when both parties have accounts with the same carrier or have cross carrier agreements. There is no inherent method for dynamically determining the servicing wireless carrier from any standard public information. To implement such features, there would have to be extensions to a contacts database that specify the hosting carrier. If the two carriers match, then a free cellular call would be possible; otherwise, a standard PSTN routed call would be required.

- *Outbound call support only.* Calls originated from the mobile handset to other phones would be easily supportable, but inbound calls could not support the peer-to-peer call scenario due to complexity in such a call setup scheme. Additionally, cellular calls between peer mobile handsets would be difficult to support because it would require the originating user to remember the cellular number of the contact rather than the extension number.

In the end, attempts to take advantage of free cellular calls would impose some real inconveniences on the caller and would not be practical. However, there may be some market demand for this level of least-cost feature support.

13.9.3 White/Black Carrier Network Lists

Hefty charges can be incurred in crossing certain carrier network boundaries. This is particularly true in Europe. Therefore, to avoid the high roaming tariffs imposed by some carriers, the UMC solution might provide a *blacklist* feature that would allow specification of specific carriers that would be blocked from use due to the incurred roaming charges. Though such a feature does have its appeal, the realized value may be questioned because of the increased TOC for managing such databases and the inconvenience of losing telephony connections when in range of such carrier services without a corresponding WiFi/Internet connection.

CHAPTER 14

Mobilizing Applications

14.1 Telephony/Email

The previous chapters of this book focused on how UMC can extend the mobile capabilities of a telephony application. For a business, the telephone is its lifeblood for communicating to associates, vendors, and customers. The other communication lifeblood of business is email. The importance of mobile email was demonstrated clearly by the meteoric growth of Research in Motion's BlackBerry product, which supports mobile email over a cellular data network. The size of this market is billions of dollars per quarter, and demand shows no sign in leveling off, with double-digit CAGR every year. What is traditionally lacking with mobile email (as with mobile telephony) is agility to operate across multiple wireless networks. There is no technical reason that email could not be serviced in the same manner as UMC telephony applications with bridged WiFi and cellular network access.

Indeed, as UMC applications enter the market, existing mobile email applications can execute in parallel but may have *dead zones* of connectivity, when cellular data services are not available. Moving ahead, market demand will drive the mobile email vendors to extend the network agility support to match that of UMC telephony capabilities. Converged UMC telephony/email solutions will emerge on the market, rivaling existing solutions and accelerating the evolution toward support for cross-wireless network access. The same cost and control requirements that drive a WiFi/cellular UMC solution for telephony will also drive delivery of a UMC mobile email solution.

14.2   Instant Messaging/SMS

Perhaps the next most important method of mobile communication today is text messaging. Traditionally, over the Internet instant messaging (IM) has been used for peer text messaging and has come to be expected as a feature supported on a networked computer where Microsoft Messenger, Yahoo Messenger, and AOL are available to almost every Internet user. Correspondingly, Short Message Service (SMS) is a peer to IM on the cellular networks. Each year, mobile cellular messaging has demonstrated strong growth. Gartner Group reported that "2.3 trillion messages will be sent across major markets worldwide in 2008, a 19.6 percent increase from the 2007 total of 1.9 trillion messages. Mobile messaging revenue across major markets will grow 15.7 percent in 2008 to $60.2 billion, up from $52 billion in 2007."[1]

With the availability of dual-mode handsets, mobile individuals face the possible use of two different messaging systems associated with two different directories. One method requires the mobile phone number for the individual contact, and the other requires the email or IM ID. Both of these data objects can be resident in a *contact* database, but unification of these two messaging methods would greatly simplify and accelerate mobilizing the ability to send a message to anyone from anywhere. Such unification will require some centralized bridging/gateway service that *could* be provided by a UMC mobility server. Some commercially available UMC solutions provide for network-agnostic IM applications.[2] Today SMS message traffic is managed in parallel with such UMC applications, but there is no commercially available unification of these messaging classes. The ideal would be the ability to send any arbitrary text message to IM, email, or SMS target (or all three). A generalized message service architecture can be easily envisioned (see Figure 14.1) but difficult to implement.

Most carriers support an email⇨SMS translation service, and some carriers and carrier-independent providers support a SMS⇨ email service. The SMS email functionality imposes an address-mapping management requirement that would be *pushed* to the end user because multiple email addresses may be desirable to receive SMS traffic. Such a mapping service must reside in the carrier *cloud*. How such services finally get integrated into UMC applications will depend on the application vendor.

[1] Gartner Group, *Market Trends: Mobile Messaging Worldwide, 2006–2011* (Gartner report, 12-17-2007).
[2] DiVitas Network's Mobile UC solution.

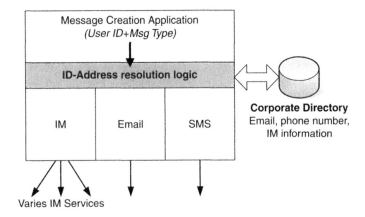

Figure 14.1: Generalized IM service layer.

14.3   Push-to-Talk

Push-to-Talk is classically an implementation of the *one-to-many* communications mode that was made famous by the World War II-era walkie-talkies. The original RF architecture was based on a set of frequencies assigned as *channels*, where each channel was a transmission media for half-duplex, one-to-many communications. Such exchanges are popular and vital for many vertical market industries such as construction, public safety, and healthcare. This communication mode is a great complementary benefit to standard UMC telephony services.

Push-to-Talk (PTT) has been a rousing success with cellular providers. Leading the market was Nextel, with a feature that allowed one user to initiate a one-to-many conversation with others who were assigned to the same channel group. Filling a different communications mode than the traditional peer-to-peer model of a telephone call, cellular PTT is now supported by most Tier 1 wireless carriers. With regard to UMC, the question arises as to how PTT gets supported when a handset is within WiFi coverage.

Today there are no commercial UMC solutions that include support PTT on either cellular or WiFi. However, as the market moves forward, demand for this feature will increase. Most likely, initial offerings will support PTT over cellular and PTT over WiFi but without transnetwork communication bridging.

14.4 Wireless Video: Monitoring and Conferencing

Once a handset that is mobile has an IP address, the application world opens up. Streaming video becomes an option that is only limited by the resolution of the screen and the end-to-end bandwidth of the network connection. The underlying architecture for support of VoIP is very similar to any architecture required to support video services, and converging mobile voice and video applications results in *mobile videoconferencing*. Some softphone vendors currently offer WiFi-based video capability, but there are no such UMC commercial offerings as of the third quarter of 2008.

Support for this level of real-time application will come to the market within the next two to five years. Adoption of higher-bandwidth, newer in-building wireless technologies (802.11n or WiMAX) and wide area wireless (UMTS) will catalyze development of these exciting new applications. Once they're in place, a whole new horizon of mobile applications will open up.

14.5 Vertical Market Opportunities

Beyond the basic horizontal applications of voice, messaging, email, and video, a host of vertical market applications are ripe for UMC support. Demand for mobile voice applications is greatest in healthcare and higher education, but as market competition and manufacturing volume drive component prices lower, new application opportunities will open up that are currently cost bound. Table 14.1 provides only an example list of vertical market opportunities that can take advantage of UMC network agility.

Table 14.1: Vertical market mobile candidates

Application	Industry	Comments/Description
Nurse call	Healthcare	Telephony applications that link patient alerts with coordinating responses from healthcare professionals.
Sales force automation (SFA)	Retail	Mobile tracking of leads and opportunities is a true value-add for real estate and others.
Customer relationship management (CRM)	All horizontal	Engaging a customer wirelessly through a CRM will be a real boon for many traveling sales teams.
Field service automation (FSA)	Service	Supporting multimedia voice and data to mobile field service personnel for problem identification and resolution.

The list of potential mobile vertical market applications is literally endless and will be fully explored as the UMC market matures.

14.6   Mobile Location-Based Services

Being untethered yet accessible by either voice or video modes poses new management and security challenges. Without an underlying location-based service, there is no way to know where the caller may be located. Some individuals may see this—to be able to meet their job responsibilities with little or no restrictions on their physical locale—as an advantage. However, location information regarding a mobile device can be very important.

Modern cellular phones support emergency services through a coordinated infrastructure that quickly identifies the user's physical location and can direct emergency response teams to that location. This can be done through mapping current cell tower information or GPS coordinates for handsets that support this technology.

Managing and mapping location information for WiFi-linked devices is more challenging. For static VoIP phones, most service providers have approached a 911 solution by having the user submit location information (usually a home address) as part of the registration process. A UMC handset, however, can be attached to the Internet anywhere in the world when an emergency occurs. There is little implicit location information that can be derived from an IP address or other network information.

Enterprises may also want to factor in location information with enforcing management policies or support of specific vertical market application functions. Knowing where a handset might be if that device has been reported stolen is also important. Equally important might be to have location information upon failure of a particular device. Location of mobile devices will evolve to become an important aspect of any UMC application, especially for mobile enterprise solutions to address asset management and security/emergency requirements.

14.6.1   Real-Time Location Services (RTLS)

To address the WiFi location management requirement, a number of vendors have implemented extended WiFi features for their wireless voice product under the term *real-time location services* (RTLS). There are basically several approaches to supporting WiFi RTLS in a UMC context:

- *WiFi pseudo-GPS.* Approximating the location of a user from access point roam history and current AP association. This method has some merit but assumes that a facility map has been generated for the placement of the APs. The resolution of this method is within several hundreds of feet, depending on the layout of the building and coverage. The one problem with this approach is that it fails to accurately support three-dimensional placement. A user, for example, might be associated with an AP on the floor above when an emergency call is placed; emergency responders might be sent to the wrong floor if location information is based solely on current AP association.

- *Soft GPS.* A more sophisticated location-based service is possible through factoring a network map, association history, association signal strength, and some triangulation calculations. One company[3] has commercialized this approach using management software and their RF *tags*, but their value-add functionality must be integrated into the final UMC solution package.

- *RTLS infrastructure.* Use of an intelligent and parallel location management service, RTLS can be provided with great accuracy. One such RTLS vendor is WhereNet (www.wherenet.com). Though limited to RTLS monitored facilities (campus coverage), such solutions bring a high degree of real-time location information for WiFi-based devices.

14.6.2 Location-Based Services (LBS)

The ultimate UMC location services will be reachable with dual-mode handsets that also support GPS. With this service enabled, a 911 call from a handset will be able to report global longitude and latitude with amazing accuracy. Beyond emergency services, GPS location tracking can provide enterprises with workflow and efficiency kinds of information, even lost-phone-tracing capabilities. This kind of service has yet to be integrated into any of the UMC solutions, but it is definitely on the horizon.

14.7 Mobile Application Summary

As with any maturing technology, as the supporting mobile infrastructure components become more standardized, the natural migration of ISVs is to the application layer. It is at the user interface layer that these mobile applications will find the greatest market opportunities. Eliminating *information retrieval* barriers is one of the major hurdles for enterprises and UMC solutions pioneering the base services for addressing this problem.

---

[3] Ekahau (www.ekahau.com), Vision-Easy RTLS software.

CHAPTER 15

The Final Challenge: Sales/Support Channel Considerations

15.1 What Drives UMC Sales?

You cannot build a successful business on the product "wow" factor[1] alone; someone has to recognize a need or problem that is addressed by this product and will want to purchase the solution. As with most disruptive technologies when they're introduced, there's a sequential adoption dynamic that goes from concept to pioneering to evangelism to homesteading to maturing of the market. In the case of UMC, many potential customers don't even know they have a problem that needs to be solved. Accepting the technological status quo lulls buyers into a state of acceptance of "what is" without seeking alternative solutions to their personal and business communication needs.

Building a UMC product is only part of the process that is required to be commercially successful in this market; a sales channel must also be defined and developed to deliver and support the product to the target market. Sales are achieved when the solution to a customer's problem is met with appropriately priced solutions. This is true for any product, anywhere.

Defining and understanding the customer's needs is central to building a channel so that the proposed product can be properly positioned and understood. In the case of UMC solutions, the customer's problem and what drives a purchase decision will differ based on whether the product is for an individual consumer or an enterprise mobile worker.

---

[1] Unless your product is the Apple iPhone, an instant market success!

15.1.1   What Drives Consumer UMC Purchases?

What are the frustrations of the average mobile consumer? His or her mobile phone is not used for business but for personal use, much like an extension to the fixed home phone. Most consumers are concerned about managing their monthly cellular minute usage so as not to exceed the contracted limit and accrue higher per-minute charges. In the modern age of mobility, this is problematic since a larger number of consumers now use their personal cell phones in place of their home phones.

Since the mobile phone is fast becoming the de facto replacement for the fixed-line home phone, having a service option to utilize a home WiFi/Internet connection (or femtocell) for servicing home calls at a lower cost is the perfect answer. Indeed, several carriers now offer "home" services at a flat rate per month for calls placed through the home Internet connection that don't impact the total cellular minute plan. For most consumers, this option is ideal and appealing. The wrinkle is to make sure that the cost to upgrade to the dual-mode phone does not impact the overall ROI of the service agreement.

Consumer-focused UMC solutions have a *cost-centric* "go-to-market" strategy[2] and will need to effectively lower the overall per month charges for calls made from home. Demand for similar services from public hotspots will be low or nonexistent.

15.1.2   What Drives Enterprise UMC Purchases?

The considerations that will drive an enterprise UMC purchase decision will be completely different from those driving individual consumers. Most enterprises have two major requirements regarding support for their mobile workforce:

* Control and management of enterprise-owned mobile devices

* Leveraging new wireless technologies to maximize mobile worker productivity

Monthly usage costs, though important, are not the highest priority in the enterprise budget priorities. If the CFO knows the aggregate monthly cellular usage of his mobile associates, he can negotiate with the carrier for a lower rate based on increasing volume. With this approach there is no simple method to collect the aggregate usage. Usually, each cellular phone user receives a hardcopy of the monthly

---

[2] Example: T-Mobile's @Home program provides unlimited WiFi-minutes for $20/month service fee.

minute usage, which would have to be manually collected and summed to produce such a report. For enterprises to manage their cellular usage and contracts, some central reporting capability should be available.

More important than managing the monthly cellular usage is improving the productivity of the mobile associate. The decision chain in any organization can be broken when a mobile decision maker is unavailable or inaccessible for any period of time. Even if this unavailability is brief, it can translate into lost business or missed opportunities (see Figure 15.1). UMC solutions for enterprises hold the promise of virtually eliminating those times of inaccessibility and allowing the key associate to make those important decisions regardless of where they may be. Leveraging the on-campus WiFi allows an enterprise to achieve a new level of accessibility on campus without adding to the cellular call plan or installing an in-building cellular network. Leveraging the cellular connection that is coupled to the company telecommunications systems allows the enterprise to extend that accessibility outside the office.

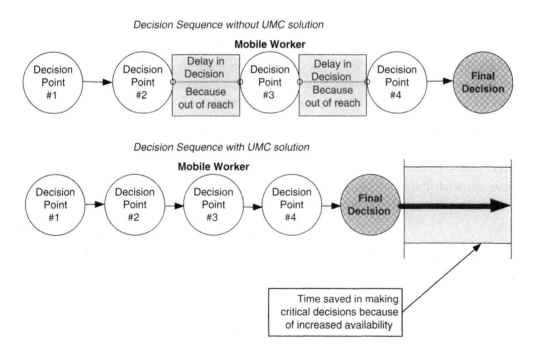

Figure 15.1: Decision sequence efficiency.

In a recent Aberdeen Group report, 230 top-line enterprises were surveyed to determine the effectiveness of FMC solutions in their organizations.[3] The study results showed, among other things, a 54% reduction in time required making a decision. For enterprises, UMC products must have a reasonable ROI, but the benefits achieved from these products go beyond simple cash accounting to maximize productivity and efficiency within the organization.

15.2   The UMC Purchase Challenge

Once the UMC elements have been integrated as a solution, the question becomes *how to deliver it to the customer*. Classically this is a *sales channel* problem, which consists of identifying how business entities find UMC prospects, install the product, and provide ongoing support through the life of the product. In the same manner that the UMC solution components are fragmented and come from a variety of vendors, an equal complexity is faced when attempting to define how an effective sales channel can be developed.

Classic sales models consist of a *tiered* approach. A manufacturer can sell *directly* to its consumers, but this means that all the responsibility for sales and support falls on this one company, which can have geographic limitations. This model works well for certain classes of products and with companies that are large enough to support national and international sales offices and have 24/7 support teams. With a UMC-class product, examples of such large companies might be Avaya or Cisco Systems.

The vast majority of UMC market players, however, will sell their product through either *distribution* or through regional and national systems integrators (SIs) or value-added resellers (VARs). This approach allows a manufacturing concern to focus on what it does best—manufacturing—and leverage the sales power of other companies already successful in selling to their prospective customers.

15.2.1   Consumer Sales Channels

UMC solutions that are targeted to the consumer market have a rather simple challenge because they can exploit sales/support channels that already exist (see Figure 15.2). In this model, it is the wireless carrier that is the source of dual-mode handsets and

---

[3] Aberdeen Group, *Fixed Mobile Convergence in the Enterprise*, March 2008.

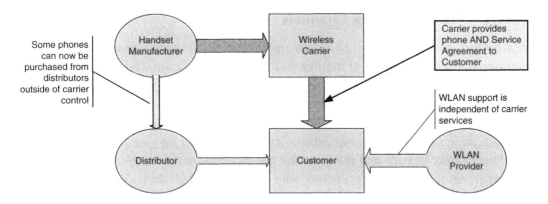

Figure 15.2: Consumer UMC channel model.

UMC mobility service agreements. A carrier deciding to support a UMC solution will upgrade its network and engage a handset partner to create a dual-mode handset conforming to the carrier-selected UMC architecture.[4] Product marketing to the consumer will be through existing television, radio, and publication ad channels and sales through existing retail outlets.

The UMC purchase options for consumers will be virtually the same as standard cellular phone purchases: procure from the carrier retail store. This single-source approach has been highly successful for the carriers and will be the major sales channel focus for the new mobility solutions. As the market evolves, however, there will be alternative sources for UMC customers to purchase the dual-mode handsets independently of the carrier. Certain regional and national distributors are now offering dual-mode handsets for sale directly to the consumer. The consumer must still engage the carrier for purchase of the SLA and must make the handset purchase decision from a list of handsets that the carrier has precertified.

Completion of the UMC wireless environment is the WLAN component. For a UMC *home* consumer, this could mean purchase of a WiFi router to connect to her Internet connection or accessing the network through a commercial WiFi hotspot service that is sponsored by the carrier.[5]

---

[4] In the majority of cases this architecture will be the UMA/GAN 3GPP approach to UMC.

[5] Example: T-Mobile's hotspot services found at Starbucks.

15.2.2   Prosumer Sales Channels

A *prosumer*, or a *professional consumer*, is typified by the road-warrior or corridor-warrior type (rarely at their desks) who are prime candidates for the enterprise-class UMC solution. These are the highly mobile executives, sales, or field support personnel who require reliable mobile communications solutions to get their jobs done. Though it is possible for a *prosumer* to purchase a *consumer* UMC product, the resulting mobile capabilities are not integrated into the company communications systems and therefore are not as effective a business tool. With consumer solutions, it is the consumer who makes the "buy" decision, but with prosumer solutions, it is the deploying enterprise management that makes such decisions; because of this purchasing dynamic change, the buy decision is typically more complicated due to consideration of cost, reliability, supportability, and viability of the supplying vendors.

Enterprises and SMB businesses typically do not purchase directly from a product manufacturer but rather deal with regional or national SIs and VARs. These *channel* businesses act as a single source for selling and supporting ancillary products and services not directly supported by the purchasing enterprise, bringing skills and specific technology knowledge to the solution absent within the enterprise. The channel dynamic that has evolved to deliver UMC-class products is a worldwide multitiered structure (see Figure 15.3).

The multitiered sales channel may contain three major contributing business classes:

- The original component manufacturer

- Regional/national distributor

- Regional/national SI or VAR

These three business entities make up the "channel" and have a symbiotic relationship in developing, installing, and supporting the end-to-end UMC solution. Though such a structure is not hard and fast in implementation, most WLAN, PBX, and network products are sold through a similar structure. It is important for the end consumer of a UMC product to have a single vendor on which to rely for all aspects of the product—one "throat to choke" for deployment and support. Such a sales/support channel did not exist until the appearance of UMC-like applications that spanned such a broad set of technologies. A full-service SI or VAR would have to have expertise in WLAN, cellular network and handset, PBX, and networking products to provide such

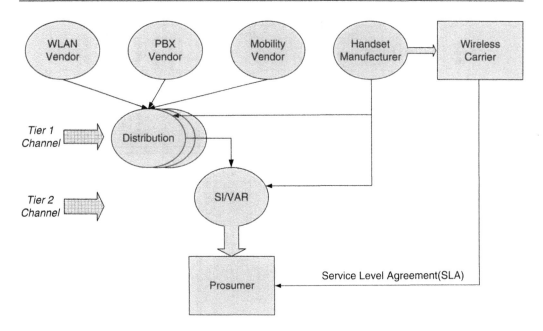

Figure 15.3: Prosumer UMC channel model.

a single-source point of sale. Most existing channels, unfortunately, are focused on only one or two of these technologies and not the full technology breadth required for a UMC product.

Enterprise-class UMC solutions will be delivered only with the occurrence of *channel convergence*, or channel partnership collaboration—a maturing of the existing channels. No one vendor sources every necessary component for a complete UMC solution. Carrier solutions are the closest to being a single vendor source but depend on consumer access to WiFi and Internet services. In the enterprise-UMC space, the competitive vendor solutions are more fragmented. Depending on their core technologies, each UMC market entrant has certain technology and/or customer dependencies to complete the full installation. Table 15.1 provides a generic overview of the completeness of each UMC competitor's solution and identifies the inherent dependencies.

Table 15.1: UMC vendor landscape

	Dual-mode Handset	UMC Client	Wireless LAN	Network Infrastructure	Cellular services	UMC Server	PBX
Network System Vendor	n/a	n/a			Customer Procured	n/a	Customer Procured
Network System Handset Vendor				Customer Procured			Customer Procured
Carrier UMC			Commercial and Home Hotspot	Internet			
Wireless LAN Vendor	Partner with handset vendor			Internet	Customer Procured		Customer Procured
PBX/Handset Vendor (1)			Customer Procured		Customer Procured		
PBX/Handset Vendor (2)	Partner with handset vendor			Customer Procured			Customer Procured
PBX Vendor (1)	Partner with handset vendor				Customer Procured		
PBX Vendor (2)	Partner with handset vendor				Customer Procured		
Handset Vendor (1)		Partner with UMC ISV			Customer Procured		Customer Procured
Handset Vendor (2)				Customer Procured		Partner with UMC ISV	Customer Procured
Enterprise UMC ISV	Partner with handset vendor			Customer Procured			Customer Procured

Several major categories of UMC vendors will be coming to the market and will include:

- Wireless carriers with a consumer-UMC solution

- PBX/telephony vendors with an enterprise-UMC solution

- Wireless LAN vendors with an enterprise-UMC solution

- Handset vendors with enterprise and consumer UMC solutions

- Independent software vendors (ISVs) with enterprise UMC solutions

15.2.3 Wireless Carrier UMC Solutions

The purpose of UMC services for a wireless carrier is to extend their available market and minimize subscriber churn through extended services. Wireless service providers have an existing sales/support channel they can leverage but have two major dependencies to address in bring the product to market. First, they must partner with major handset vendors to provide the requisite dual-mode handset products. Cellular phone manufacturers such as Nokia, Motorola, Samsung, and others have announced their commitment to UMC technologies to meet this requirement.

To address the second requirement, carriers must partner to provide access to the WiFi services. This simplest partnering is for end users to leverage their home hotspots (WiFi/Internet) as wireless service access points. For example, in the past year, T-Mobile and France Telecom have successfully launched their "at-home" services in regional pilot areas. To extend the WiFi access capabilities for these UMC products beyond the home, carriers must ally themselves with large commercial entities that provide WiFi/Internet services as a convenience or for paid service to their customers. To fully meet the UMC market and business goals of a carrier, identifying, engaging, and rolling out these partnered services will be the major hurdle. It is not likely that a commercial business will deploy a WiFi service just to support the UMC market opportunity and will have made that decision based on other business factors. Adding UMC voice traffic on top of a hotspot deployed for Internet or email applications may be problematic for all parties.

The greatest marketing hurdle for a carrier's UMC success is its need to provide sufficient WiFi service access across the geographic breadth. A secondary consideration is the importance of meeting the mobility needs of the carrier's customer base. Because of this later consideration, the expectation is that many carriers will pursue a dual strategy to support femtocells and picocells solutions.

15.2.4   PBX/Telephony UMC Solutions

The business motivation for telephony vendors to provide UMC products is completely different from that of the wireless carriers. The major PBX vendors have been in the process of shifting their products from the legacy TDM designs to a VoIP-based design over the past five or six years. The emerging market is demanding standards-based iPBX solutions; this presents a strong challenge to the PBX market, which has traditionally been a proprietary solution approach. The availability of standards-based VoIP solutions now offers the possibility of intervendor interoperability, which significantly increases the competitive challenges for each vendor. Adding the demand for mobile telephony functionality simply raises the bar on these challenges. The main challenge for legacy PBX vendors is to retain or increase their market share in the face of increased competition from traditional competitors and new pure-VoIP market entrants.

The traditional PBX market players are motivated to provide products that stabilize their customer base as the shift from TDM to VoIP telephony occurs, and all credible PBX vendors now offer pure VoIP or hybrid VoIP solutions as part of their product lines. Additionally, most offer some kind of mobility solution. In the same manner that carriers partnered with handset vendors for a solution, PBX vendors must do the same.

An initial offering from PBX vendors is typically the ability to *call forward on busy or no answer* to a cellular phone number. This is a simple extension to features they have supported for a number of years. The more aggressive PBX vendors are also offering their own UMC (a.k.a. FMC) solution by partnering with the top handset vendors. Most of these vendors support the latest generation of Nokia dual-mode phones and have made announcements about support of other OS platforms.

The risk that PBX vendors run in promoting UMC solutions is the erosion of their sales of proprietary (or vendor supplied) desktop phones. Even if these phones are based on VoIP standards such as SIP, they will extend the capabilities of these phones to provide value-add beyond simple telephony functions, and support of a rich UMC solution may minimize sales of these profitable desksets. The market is demanding UMC solutions, and PBX vendor UMC solutions are commercially available from companies like Avaya, Siemens, NEC, and Nortel.

15.2.5   Wireless LAN UMC Solutions

Several wireless LAN vendors have announced their entrance into the UMC market.[6] Other major WLAN vendors have also begun disclosing their strategy in meeting the mobility market demand.[7] The challenge for WLAN vendors is to be able to provide value-added features that are above and beyond the standards-mandated basic telephony features. In appealing to the enterprise and SMB market, the ability to support WiFi voice at any level is a competitive benefit in the current market. By tightly integrating a UMC solution with their products, these vendors have addressed several market hurdles: simplification of LAN/UMC management and deployment costs and providing a single LAN infrastructure for both voice and data. Depending on the specifics of the implementation, the only UMC component missing from the solution may be the dual-mode handset. With these solutions, the approach is usually agnostic to the wireless carrier. Partnering with a handset vendor is the lynchpin of WLAN vendor success. Depending on the UMC architecture implemented, a special handset client must be developed and may be proprietary to that vendor's solution.

Support for integrating into a business PBX may also be problematic with products from such vendors. The complexity of options for achieving such integration is tremendous and can place a burden on the field service and support organizations. The typical approach is to *certify* several PBX/iPBX products that have a significant market share and sell to that market for any launch.

15.2.6   Handset Vendor UMC Solutions

Success in the UMC market for handset vendors is almost assured no matter which go-to-market strategy they choose. Most certainly, all other market contenders will partner with them due to this requisite UMC component. The challenge for the handset vendor is which market partners, UMC architectures, and cellular technologies to support. Support for GSM, the most prevalent worldwide cellular technology, is straightforward; dual-mode phone support for CDMA networks has lagged behind. Unfortunately, this leaves much of North America short on a UMC handset solution since CDMA is still the dominant cellular service in many areas of the country. Which UMC partner to align with is also fairly simple; you partner with all of them

---

[6] Aruba Networks has announced that its embedded solution will be available in mid-2008.
[7] Cisco announced (June 2008) its Mobile Services Engine (MSE) to deliver and enable mobility solutions.

because market demand will drive such decisions. Most of the major handset manufacturers have already announced support for 3GPP and UMA but are positioned to support ISV-class solutions (see next section).

There are a few handset vendors that are unique in the market in that they are part of major electronic concerns that also provide other components for a UMC solution; they provide either network infrastructure, wireless LAN, and/or iPBX solutions. These vendors are in a unique position to come to market with a minimum of dependencies, even with enterprise-centric solutions. Most certainly companies such as Cisco, Nortel, HP, and a few others fall into this category.

The major hurdle for many handset manufacturers lies beyond addressing these solution dependencies and is more practical: What is the size of the market? Such vendors may be accustomed to producing and selling hundreds of thousands or even millions of handsets per year. The nascent UMC market may only begin the ramp-up with tens of thousands of annual worldwide sales. Such a business consideration can impact the investment a manufacturer may make in the tools and engineering necessary to meet the relatively low UMC demand. Their manufacturing systems have been optimized for production levels orders of magnitude greater than what is anticipated, and margin projections may not be enticing.

Even with these hurdles, the UMC market is set to launch, given the commitments already made by the handset vendors. The question is, "How fast will the UMC market grow?"

15.2.7   Independent Software Vendor UMC Solutions

One market segment that has attracted a number of ISVs is the enterprise UMC market. These products are in direct competition with some PBX and networking vendor UMC solutions, but such solutions are not vendor specific. The broad UMC market is fragmented with respect to PBX, wireless LAN, and cellular service deployments, which beg for an agnostic UMC solution. ISVs can approach the broad segment of the market and offer an architecture that is designed to integrate into customer sites where the mix of WLAN, PBX, and carrier requirements can be diverse.

The potential market for an ISV-agnostic solution is much larger than any single PBX or networking vendor solution, but it also has more solution dependencies. Like most other UMC solutions, there are dependencies on partnering with a dual-mode handset vendor. ISV partnering may be different than what is required in delivering other UMC solutions in that the ISV will provide its own client for the handset that provides the UMC functionality. This is due to the PBX and carrier-agnostic approach, where delivery of the application relies on management of both ends of the handset/server connection. In such cases, procurement of the target handset will be through *standard* sales channels, and the UMC client is installed at the customer facility.

Beyond the requirement of a handset solution, the next major solution hurdle is providing the integration to the PBX or iPBX. To provide a unified solution that integrates with the top global PBX and iPBX systems is a major challenge. All TDM PBXs are proprietary, and even iPBXs from some vendors may be proprietary. Broad support for this level of integration requires additional effort on the part of the ISV to "certify" its application with specific PBX/iPBX solutions. Depending on the base technology and the nature of any proprietary features, this requirement is a significant challenge.

The solution dependencies with which ISVs are faced are not insurmountable, and several such products are on the market, including solutions from DiVitas Networks, FirstHand Technologies (acquired by CounterPath in 2008), and Agito Networks. Each vendor takes a slightly different approach to the solution, but all are positioned at the enterprise/SMB market as a mobility overlay to existing telephony and network systems. The ability to extend the mobile access of an *office* PBX phone beyond the bounds of the office has a significant impact on the accessibility and, consequently, the productivity of the mobile office worker.

CHAPTER 16

Unbounded Mobile Communications: Beyond FMC

16.1 Nothing But Change in the Future

Where do we go from here? At the time of this writing, a number of first and second generations of UMC solutions (FMC, eFMC, Mobile UC, UMA, GAN, etc.) are commercially available. These solutions will be marketed by the respective companies and can meet many of today's mobile communications requirements. There are, however, major changes on the horizon regarding all the technology and service components of the UMC offerings. It's important to understand how some of these trends will impact product offerings:

- In general, the evolution of wide area networks will be moving toward more IP-centric architectures with better geographic coverage and faster data channels. This puts them in a position for support of future IMS networks, which are projected to become the backbone of future worldwide networks.

- The term *fixed/mobile convergence* will pass into history and be replaced by the more extensive term *mobile unified communications*, which covers the transport agility of FMC and the IP backbone of UC.

- Mobile communication options become more important in the context of worldwide connectivity because individuals are more mobile than in previous generations, yet still need immediate communication options. Associated nonmobile groups may also be geographically dispersed and need to be in contact between any two points or individuals on the globe. This market demand will drive more UMC opportunities.

- Service for network presence will evolve to allow network participants to not only be aware of someone's online status but also information about their location, activity, and methods of communication.

- Social networks will be strengthened and expanded by availability of the virtual ubiquitous wireless UMC solutions.

- Beyond unbounded mobile voice, mobile entertainment will become a major commercial hotbed. IP-TV, news, Internet games, and wireless music access will be incorporated into converged wireless devices, replacing the portable radios and game units of the past.

- Each of the individual technology components will evolve. Handsets will become more feature rich, wireless LAN technologies will become more pervasive with higher bandwidth, and wide area networks will become more expansive in coverage. The convenience and pervasiveness of wireless services will continue to expand and be available virtually anywhere.

Are there repercussions resulting from the shift to wireless connectivity within society in general? The PSTN will be minimized over the next 20 to 30 years, and some industry pundits[1] have even proposed the death of the desk phone. Could these also be true for the consumer? Many young adults already have only a cellular phone as their primary telephony option. Could this be true for the enterprise mobile worker or prosumer? Yes, some enterprises have already made the shift to mobile-phone-only for their workers. The critical business concern is how to keep these workers interconnected to address business processes.

16.2   Handset Evolution

The handset options offered to the UMC market will be of a higher level of functional convergence moving forward. Features such as GPS, location-based services, additional WLAN and WWAN support, expanded security, and biometrics will be included. The problem of battery life continues to plague the market but will be lessened by improved battery technology.

Multiple form factors (flip phones, candy-bar phones, ruggedized phones) will proliferate, giving the end customer the broadest possible choice of mobile device

---

[1] Unstrung, July 24, 2007 (www.unstrung.com/document.asp?doc_id=129838).

form-factors. Due to growing dependencies on such mobile devices, the market will drive advances in GUI design to further simplify their use.

16.3   Major Technology Changes

16.3.1   New Wireless LAN Options

Near term, there are additional wireless technologies for UMC consideration. The availability of 802.11n devices is appealing but poses some problems with regard to handheld design footprint, power consumption, and antenna design requirements. As WiMAX, the other new contender, becomes more dominant as a "last-mile" service, this will become a consideration for support by handheld devices, perhaps resulting in a trimode (WiFi/WiMAX/Cellular) handheld.

Who knows what may become a reality in the future, but the ultimate *converged* vision for handset would be support for Bluetooth, 802.11b/g/a/n, GSM (four frequencies), CDMA, and WiMAX.

16.3.2   IPv6—Just Around the Corner

To provide more addressability (and features), IPv6 will become the prevalent TCP/IP architecture over the next three to five years. Transitioning from IP4 to IP6 will occur with the Internet and corporate networks evolving to this standard, and the impact of delivery of UMC capabilities will be enhanced. The exact timeline for this transition is still to be worked out, but it will become a reality in the 21st century.

16.3.3   Carrier Evolution: 3G to 4G

3GPP has begun defining long-term evolution (LTE) as the future for enhancing UMTS services over wide area networks. LTE is not a standard but an initiative to create new standards for Release 8 of UMTS, part of the 4G networks supporting the All IP Network (AIPN). Significantly higher packet data rates will be supported with more efficient use of the RF frequencies. Elements of LTE will be introduced by carriers over the next three to four years.

Carrier evolution has been a constant reality. The mobile subscriber base continues to demand more than just voice. Data services for IM, surfing the Web, and running specific applications now drive the evolution of wide area wireless network providers. The transition from 2G/2.5G to 3G has not been complete in many areas, but a view to 4G is already a reality. The goal of a 4G network will be to provide a wireless packet-based solution in which high data rate voice, data and streamed multimedia can be offered without geographic bounds.

16.3.4   Antenna Options

To allow all types of wireless services to be more pervasive inside buildings or on campuses, the Distributed Antenna System (DAS) is being rolled out by some vendors. This concept is to have a single-antenna system capable of supporting multiple wireless technologies without the limitations often encountered by traditional antenna solutions. Traditionally, to deploy cellular and WiFi inside an office meant deploying two distinct antenna systems that required their own maintenance upkeep. This approach is expensive and unappealing to large corporate IT departments. Deployment of a DAS appears to answer the complexities of covering a facility with a single antenna system. With such solutions, however, there are expense considerations; they are more expensive than the single-antenna infrastructures and may be limited by vendor contract as to what services may be deployed over these networks. For example, some wireless carriers are now offering DAS solutions to enterprises, with the caveat that WiFi not utilize this antenna system.

16.3.5   VoIP Future Evolution

SIP dominates the VoIP landscape today, but new multimedia protocols are being considered by standards bodies. The International Telecommunications Union (ITU) is considering creation of a new protocol: Advanced Multimedia System (AMS). The effort is under the auspices of the ITU-T SG16 group and has a vision of creating a new protocol to provide "anywhere, anytime" connectivity. This vision is clearly aligned with the higher-level user vision of UMC. Once defined and adopted (estimated schedule: 2012), such a protocol may replace the existing H.323 and SIP protocols.

16.3.6   IMS

The goal of the IMS design is to provide a framework that can leverage all existing communication networks (PSTN, cellular networks, and Internet) to construct a set of services that will provide for real-time application voice and data integration. For example, with a mobile device, a user could simultaneously carry on a voice conversation and share data/images/files with the other parties from a single mobile device. Access to the diverse wireless networks assumes the availability of a dual-mode device (WiFi or WiMAX and/or GSM or CDMA) that is IMS aware. Borderless connectivity is a major goal of IMS.

As currently defined, IMS is an *architecture*, not a *specification*. The design embodies concepts of key "functional" components and the interconnect methods to be implemented.[2] Because the design scope of IMS is so broad, the underlying architecture is complex and has many elements defined as part of the solution (see Table 16.1).

Table 16.1: Major IMS components

IMS Component	Service	Remarks
Application Services	SIP-AS: SIP Application Services	IMS service entry point for various network applications
	OSA-SCS: Open Services Architecture Service Capability Server	Generic application platform access point to IMS services
Signaling Services	P-CSCF: Proxy Call Session Control Function	Edge service that allows an IMS node to gain access to the IMS network
	I-CSCF: Interrogating Call Session Control Function	Determines which S-CSCF to register for basic services
	S-CSCF: Services Call Control Function	The core service element for the IMS system manages the state of all the current activity
	MGCF: Media Gateway Control Function	The IMS control function that manages media gateway services
	BGCF: Breakout Gateway Control Function	The IMS control function that makes media routing decisions for egress and access to the IMS network
Media Services	MGS: Media Gateway Service	The media gateway handling the RTP streams for a session
	MRFP: Media Resource Function Processor	Can provide media management services for merging streams for conferencing and other functions
Database Services	HSS: Home Subscriber Service	Subscriber database
	SLR: Subscriber Locator Resource	Service provided to search and identify registered users

---

[2] http://en.wikipedia.org/wiki/IP_Multimedia_Subsystem.

Specifics of an implementation of this architecture, however, are left to individual providers and may be implemented in a somewhat vendor-specific manner. Therefore, it is projected that as IMS solutions are brought to market, some *certification* of interoperability will be necessitated.

To aid in understanding the morass of IMS acronyms, a little tutorial is appropriate. Many of the IMS discrete elements are labeled with the characters *CF*, which stand for Control Function. Each of these elements provides a specific capability within the IMS network. The basic groups of capabilities fall into four major functional categories:

- *Application services.* Elements that interface with the end user through IMS-enabled terminals that provide end-user application features.

- *Signaling services.* Elements responsible for setting up and maintaining logical session connections between IMS entities.

- *Media services.* The IMS nodes responsible for transport of data media. This can be data packets, audio streams from phone calls, or video streams.

- *Database services.* Repository of configuration and user information necessary to support the IMS authentication and transport services.

When IMS services are rolled out by provider companies, not all these IMS elements are required. For that reason, IMS services will most likely be rolled out in a stepwise fashion: supporting a small set of transport capabilities with the initial release followed by functional upgrades as the market grows.

Where does UMC fit with IMS's future? There are several options available for UMC systems to integrate with emerging IMS services. The most straightforward approach would be to implement the UMC components as an Application Server (AS) that uses the IMS network for basic transport services (see Figure 16.1). This method of IMS support would allow pre-IMS purchases to migrate into an IMS world, preserving the original investment.

As IMS evolves, the basic functions of UMC (consumer and prosumer) may be decoupled and diffused into the IMS cloud. With this option, some of the core features of UMC-like seamless roaming across diverse networks might be serviced within the IMS network. Customer-specific features such as integration into a corporate communication network or support of specific applications would be implemented as complementary AS modules. None of these options are available today, and they will

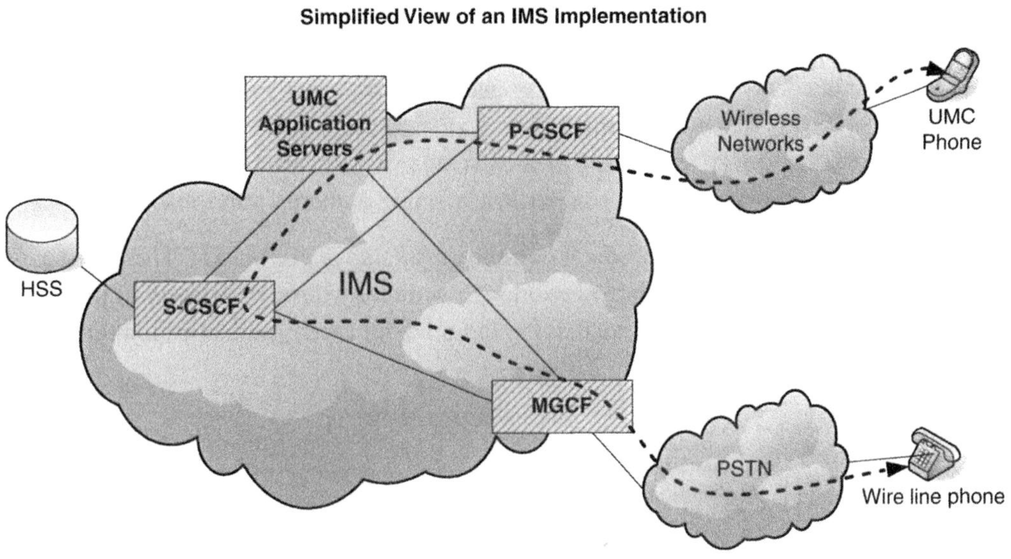

Simplified View of an IMS Implementation

Figure 16.1: IMS overview.

not appear until the IMS market matures to justify the development and purchasing expenses. The topic of IMS is immense and could fill a large volume all by itself.

Much has been written in the press regarding IMS over the past few years, and several major networking and wireless carriers have made announcements on their implementation of IMS, including Nortel, Siemens, and Ericsson. How these services will be implemented and become integrated into communication programs offered to the consumer and enterprises is yet to be seen. The early adopters of IMS will, most likely, be the smaller Tier 2 and 3 wireless carriers and PBX/network infrastructure providers. Their aggressiveness is motivated by their need to increase market share over the major players.

16.3.7 Identity Services

Support of mobile workers has expanded the need for UMC user and application identity validation capabilities. Remote users accessing critical corporate data escalate the security requirements for user identification and authorization at multiple OSI layers. Authorization and authentication of individual users have traditionally been the responsibility of each functional layer within the connection stack, but this has resulted in a management nightmare with moves-adds-changes at each layer within the logical OSI stack. The concept of *identity services* is emerging to provide a single service point that is a *pan-network* service element and can meet access control requirements at multiple levels (see Figure 16.2). Such a service, decoupled from the individual layer entities requiring authentication, can greatly simplify the task of managing on and off campus authenticated network and application access. To deliver such "engines," a *convergence* of standards must occur to describe how all these independent components may utilize a single authentication resource.

There are a few *identity-service* products currently available on the market. These will have greater impact on the overall mobile networking community through delivery of products being defined by standards bodies such as the Initiative for Open Authentication (OATH).[3]

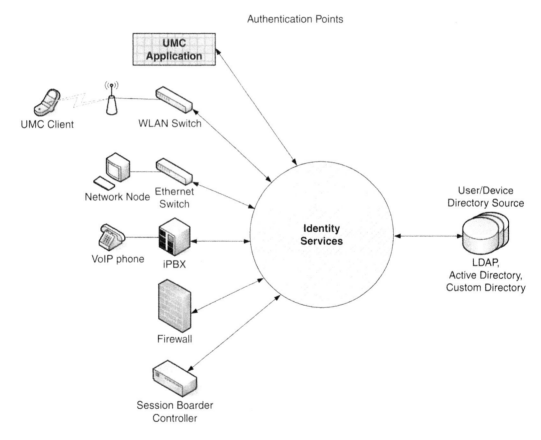

Figure 16.2: Identification services.

---

[3] www.openauthentication.org.

16.4 Presence in the Future

Reporting "presence" for a network user was incubated in the IM market. Why send an IM to a buddy if they are not there? This concept has been extended beyond its original function set and now contributes to the rich functionality required by many other application classes. For presence to reach its full market potential in the enterprise, several paradigm shifts must be made in the implementation designs of network-based applications:

- Presence-based content must be expanded to encompass more user-, environment-, and user-associated application-specific information.

- Support of presence must be decoupled from any one application and become a base service in and of itself.

- Presence must be an embedded service at the network level, independent of vendor-specific components.

- Presence should be a portable concept across any network or subsystem (WLAN, Cellular, IMS, etc.).

- Concepts of "public" and "private" presence will be implemented.

- Concept of "functional" (not individual) *presentity* could be created.

- Concept of authorized levels of presence information could be defined.

A broader definition of presence could include concepts such as:

1. I'm here (basic presence).

2. Here's *who* I am (title, skill set, responsibilities, organization position).

3. Here's *how* I can be contacted (telephone, IM, SMS or email)

4. Here's *what* I can do (capabilities and applications).

5. Here's *when* I can do it (availability).

6. Here's *where* I am (locality and time zone).

The IETF standards body has taken up the challenge of expanding the functionality of presence services and defined a Rich Presence Information Data (RPID) format that expands the concept of presence to include:

- Activities: What the person is currently doing

- Class of user: Specific user grouping for management and control

- Device-ID: Specifics about the user's device and its capabilities

- Location information: Where the user is currently located

- Location type: Office, school, church, golf course, etc.

- Class of service availability: How best to communicate with the *presentity*

- Other data descriptors, including examples such as mood of the user

Additional potential RPID parameters might include the application or communication modes available to this user and any application-specific requirements for these modes. Additional presence information could also describe the organization position, skill, and area of responsibility within an organization (i.e., chief architect, CFO, support, etc.).

16.5   Major Vendor Trends

The major networking vendors have already announced their intent to address unified communications (UC) with solutions that center around their product families. With UC, it is possible to support all kinds of communication classes that have been traditionally segregated into separate product offerings. Telephony was not integrated with data networking, which was not integrated with video technologies. UC offers new opportunities for major solution vendors to add tremendous value-add to their existing product families. UC acts as that crystallization point within the enterprise network.

Microsoft Office Communications Server (OCS), Cisco Unified Communications, and IBM Lotus Sametime are all targeted at maximizing the UC features to be leveraged from their respective core products. Presence, peer telephony, IM, and collaborative capabilities are all offered by these companies to their customer base. Each brings a competitive value distinction that enhances and extends the UC capabilities. However, none of these commercially available systems supports a mobility element that matches the UMC capabilities, a fact that will open up new markets for the UMC vendors targeting the enterprise segment.

16.6   UMC: FMC and Beyond

UMC solutions are currently on the market and meet many customer mobile communication needs. Whether consumer (controlled by the cellular provider) or enterprise (controlled by the enterprise), all the UMC solutions seek to provide an *unbounded* experience with telephony providing a consistent feature set regardless of physical location, freeing the user from concerns of coverage, location, or features.

The evolution of UMC solutions has gone quickly beyond the simple goal of providing a seamless bridge between the fixed and mobile telecommunication worlds, and the future holds promise of being able to carry a personal and professional *communicator* to be available at all times, regardless of where you are on the globe. Analysts have projected that by 2012 there will be some 5.8 billion cellular phones in use on the planet, many of which will be dual-mode, providing that extended wireless coverage needed by an ever-growing mobile community of travelers, workers, vendors, customers, associates, family, and friends.

Wireless infrastructures that blanket the globe will be the backbone of the worldwide UMC services. Once these wireless "highways" are in place, users with multimode devices will be able to maintain constant communication contact with almost anyone else on the planet. This level of technology will permit us to communicate at will with individuals and groups, without bounds.

Glossary

1XRTT

3G — Third Generation

3GPP — 3rd Generation Partnership Project

802.11 (a, b, g, n) — Four different Wireless LAN standards based on 2.4GHz and 5.2Ghz RF frequencies

AAA — Authentication, Authorization, & Accounting

AES — Advanced Encryption Standard

AIPN — All IP Network

AMPS — Advanced Mobile Phone Systems

AMS — Advanced Multimedia System

ARPU — Average Revenue per User

BSS — WiFI Basic Service Set

BSSID — Basic Service Set ID (WLAN)

CAGR — Compound Annual Growth Rate

CALEA — Communications Assistance for Law Enforcement Act

Capex — Capital Expenditure

CBC — Cellular Broadband Convergence

CCMP — Cipher Block Chaining Message Authentication Code Protocol

CCX — Cisco Compatible Extensions

CDC — Cellular Data Channel

CDMA — Carrier Detect Multiple Access

CDPD — Cellular Digital Packet Data

CDR — Call Detail Records

CFO — Chief Financial Officer

CIO — Chief Information Officer

CO — Central Office

Codec — Coder/decoder

CPE — Customer Premise Equipment

CSMD/CD — Carrier Sense Multiple Detect/Collision Detection

DECT — Digitally Enhanced Cordless Telephony

DFS — Dynamic Frequency Selection

DID — Direct Inward Dialing

DiffServ — Differentiated Services

distributed antenna system (DAS)

DMZ — Demilitarized Zone

DSCP — Differentiated Service Code Point

DSL — Digital Subscriber Line

DTMF — Dual Tone Multiple Frequency

EDGE — Enhanced Data rates for GSM Evolution

Enhanced Wireless Consortium (EWC)

ESN — Electronic Serial Number

ESSID — Extended Service Set ID (WLAN)

EV-DO — Evolution-Data Optimized (packet data service for WWAN)

FDM — frequency domain multiplexing

FEC — Forwarding Equivalence class

FEMTOCELL/Microcell

FMC — Fixed/Mobile Convergence

FMCA — Fixed/Mobile Convergence Alliance

GAN — General Access Network

GPRS — General Packet Radio Services

GPS — Global Positioning System

GSM — Global System for Mobile Communications

HA — High Availability

HDTV — High Definition TV

HIPAA — Health Insurance Portability and Accountability Act

HLR — Home Location Register

HSPA — High Speed Packet Access

ICSA — (IMS-Controlled with Static Anchoring)

IEEE — Institute of Electrical and Electronics Engineers

IETF — Internet Engineering Task Force

IM — Instant Messaging

IMEI — International Mobile Equipment Identity

IMS — IP Multimedia Subsystem

iPBX — IP Private Branch Exchange

IPSEC — IP Security

ISM — Industrial Scientific & Manufacturing

ISP — Internet Service Provider

ISV — Independent Software Vendor

ITU — Internaltional Telecommunications Union

LBS — Location Based Services

LCD — Liquid Crystal Display

LTE — Long Term Evolution

MDR — Mobile Detail Records

MIH — Media Independent Handover

MIMO — Multiple-input/Multiple-output

MMC — Mobile-to-Mobile Convergence

Mobile Unified Communications

Mobile-to-Mobile Convergence (MMC)

MOH — Music on hold

MPLS — Multiple Protocol Label Switching

MSC — Mobile Switching Center

MVNO — Mobile Virtual Network Operator

MWI — Message Waiting Indicator

NAT — Network Address Translation

OFDM — Orthognal Frequency Domain Multiplexing

Opex — Operational Expenditure

OSI — Open Systems Interconnection

OTA — Over the Air

PBX — Private Branch Exchange

PDA — Personal Digital Assistant

PRC — Peoples Republic of China

PSAP — Public Safety Answering Point

PSK — Private Shared Key

PSTN — Public Switched Telephone Network

PTT — Push to talk

QoS — Quality of Service

QWERTY — Keyboard layout form

RF — Radio Frequency

RFID — Radio Frequency Identification

RIM — Research in Motion
ROI — Return on Investment
RPID — Reich Presence Information Data
RSSI — Receive Signal Strength Indicator
RSVP — Reservation Protocol
RTLS — Real Time Location Services
RTP — Real Time Protocol
SBC — Session Boarder Controller
SDP — Session Description Protocol
SI — System Integrator
SIM — Subscriber Identification Module
SIP — Session Initiation Protocol
SLA — Service Level Agreement
SMB — Small/Medium Business
SMS — Short Message Service
SOHO — Small Office/Home Office
SRTCP — Secure Real Time Control Protocol
SRTP — Secure Real Time Protocol
SSL — Secure Socket Layer
SVP — SpectraLink Voice Priority
TCO — Total cost of ownership
TDM — time domain multiplexing
TKIP — Temporal Key Integrity Protocol
TOS — Type of Service
TPC — Transmit Power Control
U-APSD — Unscheduled Automatic Power Save Delivery
UMA — Universal Mobile Access
UMTS — universal mobile telcommunications system
UNC — UMA Network Controller
UC — Unified Communications
USD — United States Dollars
VAR — Value Added Reseller
VCC — Voice Call Continuity
VHT — Very High Throughput
VLAN — Virtual LAN
VLR — Visiting Location Register
VoIP — Voice over IP

VPN — Virtual Private Network

WAPI — Wired Authentication and Privacy Infrastructure

WCDMA — Wideband Code Division Multiple Access

WEP — Wired Equivalent Privacy

WFA — WiFi Alliance

WiFi (or Wi-Fi) — Wireless Fidelity

WiMAX — Worldwide Interoperability for Microwave Access

WISP — Wireless Internet Service Provider

WLAN — Wireless LAN

WMM — Wireless Multimedia

WPA — Wireless Protected Access

WPS — WiFi Protected Setup

WWAN — Wireless Wide Area Network

Index

Printed and bound by CPI Group (UK) Ltd, Croydon, CR0 4YY

03/10/2024

01040336-0007